THE
FOREVER
WITNESS

THE
FOREVER
WITNESS

How DNA and Genealogy
Solved a Cold Case Double Murder

Edward Humes

DUTTON

DUTTON

An imprint of Penguin Random House LLC
penguinrandomhouse.com

Images on pages 53, 115, 128, and 342 are courtesy of the Snohomish County Sheriff
Image on page 279 is courtesy of the Snohomish County Superior Court
Images on pages 166, 167, and 173 are courtesy of Parabon NanoLabs
Image on page 331 is courtesy of John Van Cuylenborg

LIBRARY OF CONGRESS CATALOGING-IN-PUBLICATION DATA
Names: Humes, Edward, author.
Title: The forever witness : how DNA and genealogy solved a cold case double murder /
Edward Humes.
Description: [New York] : Dutton, [2022] | Includes bibliographical references and index. |
Identifiers: LCCN 2021058094 (print) | LCCN 2021058095 (ebook) |
ISBN 9781524746278 (hardcover) | ISBN 9781524746292 (ebook)
Subjects: LCSH: Criminal investigation—Case studies. | Genetic Genealogy—Case studies. |
Cold cases (Criminal investigation)—Case studies.
Classification: LCC HV8073 .H877 2022 (print) | LCC HV8073 (ebook) |
DDC 363.25—dc23/eng/20220131
LC record available at https://lccn.loc.gov/2021058094
LC ebook record available at https://lccn.loc.gov/2021058095

Printed in the United States of America

3rd Printing

BOOK DESIGN BY TIFFANY ESTREICHER

*To the family and friends of
Tanya Van Cuylenborg and Jay Cook,
and to the others still awaiting answers*

CONTENTS

CONTENTS

THE
FOREVER
WITNESS

Litterbug

May 18, 2018
Snohomish County, Washington

The sixtyish man with the plain gray suit and pale blue watchful eyes
had just finished lunch when his phone buzzed. Feeling the buy-one-
get-one-free roast beef sandwiches leaden in his belly, he sighed, sure
this would be yet another false alarm. He dug the vibrating cell from
his pocket.

"Detective Scharf, sheriff's department."

"We got it!" the voice on the other end said.

Jim Scharf felt a second of incomprehension. Then the detective registered the exultant tone and who it belonged to—one of the ten undercover cops assigned to his stakeout. *Finally!* More than a week had gone
by with nothing to show for the mission but overtime bills and impatient department brass.

"What did you get?" he asked.

"His coffee cup."

Scharf paused, letting the words sink in as he sat in his department-
issue Ford SUV, stocked with enough bottled water, beef jerky, and

Arby's coupons to wait out nuclear winter, much less a long stakeout. Here he was, trying to solve one of the most baffling crimes in Pacific Northwest history, the disappearance and murder of a young Canadian couple on an overnight trip to Seattle. And now he had a coffee cup. He took a deep breath and started his car.

"Bring it to the office," Scharf said. "I'll meet you there."

THE BRUTALITY AND randomness of the murders of Tanya Van Cuylenborg and Jay Cook sparked an international manhunt, blanket media coverage, and deep anxiety that the rapist and killer might strike again in this semirural county north of Seattle. But that was thirty-one years ago. Since then there had been no eyewitnesses, no leads in the physical evidence, no arrests. Over time the panic and the headlines faded, and the investigation stalled.

The case eventually landed on Scharf's desk in the Snohomish County Sheriff's cold case unit, where his job was to bring fresh eyes to old files. Many saw this as a departmental backwater, but so far Scharf had cracked eight murders and a sexual assault no one else could solve. And now he had a surprising new lead in the Canadian double murder case.

His unlikely source was a self-taught genetic genealogist and cast member on the PBS reality series *Finding Your Roots with Henry Louis Gates, Jr.* Every cop ever assigned to the case, plus the FBI, Interpol, and even the Canadian Mounties, had all failed in their search for the killer. So this TV personality had succeeded by *not* searching for him. Instead, she had constructed his family tree.

Now it fell to Scharf to determine if she was right, and if a Seattle trucker named William Earl Talbott II really was the killer who eluded capture for three decades.

. . .

A BALDING MOUNTAIN of a man, Bill Talbott had reached his fifty-fifth birthday with no criminal convictions on his record and no known connection to the victims. Scharf assigned the surveillance detail to shadow Talbott's big rig on his daily deliveries of machine parts around Seattle, then trail him home from work, looking for anything suspicious. But other than bouts of fist shaking and shouting at other drivers, the man was a cipher. He worked, then went home and did little else.

The man's reclusiveness also made it hard for the surveillance officers to accomplish Scharf's other directive: grab something with Talbott's DNA on it without tipping him off. But there was a problem. The man never left anything behind. He was obsessive about it. And so a week of waiting and frustration went by before the break finally came where they least expected it: on a busy highway in the middle of traffic.

Talbott stopped his truck at a red light, then abruptly flung open his door and climbed onto the running board. His surprised watchers quickly slumped in their car seats, but he wasn't looking their way. His broad face florid, brows knit, Talbott leaned his bulk between cab and trailer and wrestled with something, maybe a loose cable that had been rattling and annoying him. When the light turned green, he hastily clambered back behind the wheel, and that's when it happened: a used paper coffee cup tumbled out of the cab and fell to the street below. Talbott didn't notice it, or, if he did, he ignored it and left the cup where it lay to be flattened by a hundred passing cars and trucks. He slammed his big rig into gear and roared off.

The closest undercover cop had seen it all. Leaping from his watch car and dodging traffic, the officer snatched the cup and hoisted his paper trophy high.

. . .

"I'LL DRIVE IT to the crime lab myself," Scharf told the surveillance officer when they met that afternoon at the cold case office. Ten minutes later, their paperwork signed, he was in his car and headed to the Washington State Patrol Crime Laboratory Division, the stained white cup in its plastic evidence bag on the seat beside him.

The crime lab people loved coffee cups, Scharf knew. Saliva was a rich source of DNA, and drinking always left plenty behind. Every day, we all unthinkingly throw away a plethora of objects containing our entire genome, our most private information, the road map to who and what we are. Through accident or design, Scharf didn't know which, Talbott somehow avoided doing so while under constant surveillance. But finally that cup slipped through.

"Give me twenty-four hours," Scharf's favorite lab tech told him, "and I'll have an answer for you."

The detective returned to his office with its piles of papers and yellowing photos of cold case victims, feeling anxious and steeling himself against hoping too much. He had lost count of the other suspects in this case he had sought out, checked out, and ultimately ruled out because their DNA didn't match traces left behind by the killer. Maybe this time would be different.

Tomorrow he would know. He would know the truth about Bill Talbott. He would know if he was on the verge of a big arrest and something quite possibly historic: the first-ever genetic genealogy murder trial. And he'd know if the slaying of a young couple could be solved through a family tree, a paper cup, and the heedless act of a litterbug.

Scharf flipped through a folder, its graying manila cover worn thin and soft as old leather. His case file held only one picture of Jay Cook and Tanya Van Cuylenborg together. It wasn't posed, just a hasty snapshot. Jay stood, gazing down in concentration at some small object in his hands, maybe something he was trying to unknot, his dark brown hair falling over his eyes. Tanya was seated and had just glanced up as the shutter clicked, so she appeared to look directly into Scharf's eyes. Her slightly freckled face looked relaxed, Scharf thought, carefree. He guessed it had been taken just a month or two before they died. Tanya had turned eighteen that spring. Jay was a month shy of twenty-one. They'd be pushing fifty by now, he thought, had they lived.

After a long moment, Scharf closed the folder, and waited.

PART I

WE'LL BE BACK TOMORROW NIGHT

If ever there was time, I would lie in the sands of Gonzales and wiggle my toes in the sun. So leave me not forever, and keep love in your soul.

—Tanya Van Cuylenborg,
from the last entry in her notebook

We were only eighteen. We were kids. We felt invincible. But you grow up fast when your best friend is kidnapped and murdered.

—May Robson

1

If Only . . .

November 17, 1987
Vancouver Island, British Columbia, Canada

"Why don't you come with us?" Tanya Van Cuylenborg grinned, picturing the look on her best friend's face.

"What, me?" May Robson sputtered. "Jump in the car and drive to Seattle tomorrow, just like that?"

"Nope," Tanya said. After a beat, she added, "We're going in Jay's dad's van."

This last-minute trip to Seattle was actually her boyfriend's idea, she told May. And, okay, yes, she admitted, she was feeling a little nervous about it, this first extended trip alone with Jay Cook. So Tanya was doing what any eighteen-year-old almost-adult would do in such a situation: she asked her girlfriend to come along.

But May had gone quiet.

"C'mon, Mary-Anne," Tanya implored, emphasizing each syllable of May's proper first name. "You need a little adventure."

"So true," May muttered grumpily.

Tanya held her breath. Impulsive last-minute trips abroad were not

May's thing. That was Tanya's role in this friendship. But once persuaded to jump in, Tanya knew, no one had her back better than May.

"Please come," Tanya pressed. "We're going to be sleeping in his dad's van and I'll be uncomfortable alone with Jay. It'll be so much easier with you there. It'll be fun."

Normally, Tanya could expect May to agree then, no further discussion needed. The two had done everything together, after all, ever since they bonded at a Brownie troop meeting when they were eight. They'd become a constant presence at each other's homes and tables ever since. May considered Tanya's dad, Bill, a second, funnier father. The best friends graduated from high school together, ate and drank their way through London and Paris on a school trip that spring together, bluffed their underage selves into bars together. Tanya relished making it her personal mission to coax her more conventional friend into impetuous day trips and expeditions. She couldn't remember May ever expressing regret at going along for the ride, not even when they got hopelessly lost in France late one evening. Tanya kept a cool head that night for May's sake, and they finally found their way back to their hotel, arm in arm.

This time, though, May said she'd have to disappoint her friend: she was sick. The pressure of the hard plastic telephone was making her relentless earache worse, she told Tanya. She had a pounding headache, fever, and chills. She wanted to say yes, but as much as she loved to be spontaneous with her best friend, and as much as she hated to say no when Tanya played the girlfriend-in-need card, May said she wasn't going anywhere anytime soon. Certainly not the next day.

"I'm sorry, sweetie. I'd be miserable, and I'd make the both of you miserable. I need to stay home in bed."

Tanya grumbled a bit, then mastered her disappointment, straining to sound both sincere and sympathetic when she said that, of course, she understood. Everything would be fine in Seattle, she assured May.

They both knew Jay was a great guy, that she'd be fine without May playing third wheel. She was just being silly.

In truth, she said, the impetus for the trip hadn't been fun and adventure but a request from Jay's dad, Gordon Cook. He needed a replacement furnace for a customer of his heating service and repair business. His regular supplier in Vancouver on the mainland had fallen through, but a company in Seattle had the right furnace and fittings. He just needed someone to make the five-hour car and ferry trip to pick up the old-style oil burner and haul it back.

Normally Gordon's business partner, Spud Talbot, would do it. When a trip to Seattle was needed, Spud would depart on a Friday and bring his wife. They'd make a weekend getaway of it, then return with the parts on the following Monday. But this job couldn't wait for the weekend—winter had come and the customer needed a working furnace as soon as possible.

So Jay volunteered to handle it, then asked Tanya to join him. His dad had given him money for a hotel, but Jay wanted to pocket the cash and spend the night in the family van, parked outside the supply warehouse. He and Tanya could pick up the furnace first thing in the morning, then there'd be time and money for some sightseeing and shopping before returning to Vancouver Island. They'd arrive back home that evening, Tanya explained.

"I'll talk to you then," Tanya and May said at the same time. The two friends laughed their good-nights before hanging up.

MAY ROBSON WOULD replay that phone call in her mind time and again over the days, years, and decades to come. Sometimes she would dream about it. Thirty-plus years later, she still remembers her last conversation with Tanya with the sort of clarity normally reserved for a favorite song or a beloved movie replayed more times than can be counted. The warmth of her friend's voice still rings in her ears, the laugh she knew better than her own, the easy intimacy with the one person she could and did tell everything. Except, on that last day, she told her friend no. After all this time, May still cannot speak of that without squeezing her eyes shut.

In her dream version of that conversation, May usually gives a different answer. She sees herself happily packing a bag with her mom's help, then waiting for Tanya to swing by with Jay at the wheel of his hulking copper-colored family van.

If only that dream version were true, May often thinks, everything might have been different. Two people headed to Seattle had been an easy target for a predator, she reasons. But had she gone along, had she been in that van during that trip, three might have been a crowd, and the predator might have moved on in search of easier prey. There's a chance nothing would have happened if she just had gone along for the ride, a chance Tanya would have come home as planned.

The dream, which she has had many times, evokes in May's imagination the life that might have been: Her kids would be in school with Tanya's. She and Tanya would call each other on the phone every day. Tanya would still be talking her into crazy, impulsive adventures. She had no doubt their friendship was one of those special ones that would have endured longer than anything else in their lives.

Of course, every dream has its nightmare alternative: May could have gone to Seattle with Tanya and Jay and shared their fate. A seasoned predator might not have been deterred by one additional teenaged girl. Another young woman in the van might have provided even more inducement for the killer to strike.

It's not that May hasn't thought about that chilling alternative outcome: it would have been horrible, she knows. May has lived a good life since then, and, all these years later, she is glad she chose to stay home, glad she chose life and a family and a future, even if she had not known she was choosing them at the time. But the part of her that she usually keeps to herself knows if only she had said yes to Tanya, one way or the other she would have been spared a lifetime of grief, survivor's guilt, and regret.

2

Squirrelly

Vancouver Island is Canada's warmest place, the only part of that vast country with an officially Mediterranean clime. Oak Bay and Saanich—the two beach towns where Jay Cook and Tanya Van Cuylenborg lived, attended school, and launched their overnight trip to Seattle—are side-by-side suburbs on the periphery of the provincial capital city of Victoria. Spacious brick houses line their wide streets, a canopy of maples, oaks, and birches arcing over the sidewalks and parked cars, cool green in summer; fiery red, orange, and gold in fall. Oak Bay's town motto is the Latin phrase *Sub quercu felicitas*: "Happiness under the oaks."

Drawn by the ocean views, scenic cityscape, and favorable climate in which even the chill of November feels mild, American and British filmmakers declared Oak Bay "Hollywood North" in the 1930s. For the decade leading up to World War II, locals grew accustomed to film crews and street sightings of such movie superstars of the day as Paul Muni and the first lady of American cinema, Lillian Gish. A pervasive

British influence outlasted the Hollywood years, however, and is visible in the local culture, architecture, and ivy-draped garden homes. In 1987, tea shops still outnumbered coffeehouses in the capital area (Starbucks, a small company with a handful of shops at the time, wouldn't appear in Victoria for another seven years). Visitors driving up from south of the border might well have felt they stumbled on an idealized Pacific Coast version of a New England beach town. Only when they noticed the speed limit signs in kilometers per hour and the service station marquees pricing petrol by the liter would it become clear that this definitely was not the United States—if the local pronunciation of "mum" for "mom" and "a-boat" for "about" hadn't already done the trick.

The rhythm of change here was slow, although that didn't keep longtime residents from fulminating about the loss of the good old days. Locals habitually fretted about rising crime rates in the metropolitan area of Victoria, which had just over a quarter million residents in 1987 (it would near four hundred thousand by the 2020s). But if that metro area were suddenly transported across the twenty-two miles of the Strait of Juan de Fuca and replanted on Washington's Olympic Peninsula, it would instantly become one of the safest (and least armed) places to live of its size anywhere in the United States.

Tanya and Jay grew up leaving their house keys at home. Jay's mother, Lee, was notorious for forgetting even to close the front door when she left for the day, much less lock up their Oak Bay house. Jay and his younger sister, Laura, came home one day and found the door wide open, a small stack of cash on the dining room table with a note explaining that Lee had gone to visit her family home on nearby Salt Spring Island for a couple of days. The cash, she wrote, was to cover food and other expenses while she was gone.

Nothing was disturbed or taken during the hours when the house was open and unattended. It was that kind of town, a level of community trust that was second nature to Tanya and Jay, though such feelings had long since vanished in most places in the United States.

It was about to vanish from Saanich and Oak Bay, too. Neither family would leave their doors unlocked after November 19, 1987.

EVERY GOOD RESTAURANT has a server or two who are so competent, so attentive, or so funny that regulars ask if they can be seated at one of their tables. At Pickwicks Restaurant, one such server was Tanya Van Cuylenborg, always ready to make a menu recommendation (or whisper a wry warning), share her latest comical take on affluent Oak Bay pretensions, or conjure an extra bit of dessert to distract a bored child so the parents could linger a little longer. The teenager seemed as happily quirky as the British-themed restaurant itself, an odd Oak Bay landmark that featured an elaborate clock tower populated by life-sized animatronic Dickens characters.

Pet owners in Tanya's Saanich neighborhood were equally devoted to her, their dog walker and pet sitter of choice starting at age twelve. They had seen how she trained and treated her own dog, Tessa, whose admission to the family came only after her relentless lobbying and firm promise to her dad that she would take full responsibility for the golden retriever's care and feeding. Her family had long assumed that Tanya's affinity for animals would eventually translate into a career, though she made it clear when she graduated from high school that she was nowhere near ready to make such decisions. "I just turned eighteen," she told her parents. "I don't know what I want to do next month, much less the rest of my life."

This was fine with Willem and Jean Van Cuylenborg. They were prosperous, content, and in no rush to see their youngest leave their comfortable Vancouver Island nest. Tanya had remained close to her mum and dad throughout her high school years. She could talk to them about anything, something few of their friends said of their own teenagers. Maybe that was why Tanya and her friends usually congregated at the Van Cuylenborg place rather than their own homes. Jean fed them and made them welcome, and Bill, with his prematurely silver hair and slight Dutch accent, made them laugh with improbable stories about his childhood in the Netherlands and his first jobs pumping gas and crewing fishing boats before he decided to go to law school.

"You two are so alike," May told Tanya, whose mock outrage at her friend's assertion fooled no one. May had nailed it: you could see it in their faces, their gestures and body language, even the way father and daughter argued so often and with such relish, more play and entertainment than real conflict. He was a lawyer, and she the lawyer's daughter, so debate came naturally to them, whether disputing politics or music or the quality of her tennis serve. It was how they told each other "I love you."

Lately he had noticed her thinking had matured, her arguments had grown deeper and more thoughtful. It's all coming together for her now, he told his wife not long before the trip to Seattle, and Jean could hear that tinge of sadness mixed with the pride Bill felt in seeing his girl had become a young woman. All the more reason to enjoy those lazy Sunday mornings while they lasted when, even at eighteen, Tanya would still crawl into bed with her parents as she had done since she was little.

Tanya, whose name rhymes with *can*-ya rather than *khan*-ya, was known by friends and schoolmates for her wit and her sharp tongue with those she considered fools, a trait for speaking her mind that dated all the way back to grade school. She and her childhood friend, Stephanie Krohn, would spend hours people watching in the schoolyard, at the park, or by a city fountain, where Tanya's forte was making up fictional biographies and comical internal dialogues for passersby until the two girls convulsed with laughter.

"Even then she was the ringleader," Krohn recalls. "She was never one to follow the pack. That's why it was so much fun being with her."

A natural athlete, Tanya played varsity basketball and was an accomplished tennis player, something of a family requirement given that her dad built a tennis court on the grounds of their home when she was a child. Her persona as a rebel against convention grew more pronounced in high school. In the seventies and eighties era of big hair, most snapshots of Tanya in her teens show her darkly blond hair kept short and simple. She chose clothing more for comfort, hiking, or biking than for fashion, more interested in the emerging Northwest grunge culture than the preppy look favored at her school. When most of her peers in the Oak Bay High School graduating class were out shopping for dresses and suits to wear to commencement, she told May, "No way. I'm not doing that. I'm not wearing a dress." Then she showed up to graduation in an unforgettable tuxedo off the men's rack, the jacket and pants black, with a white shirt and an eye-popping emerald-green bow tie around her neck. People couldn't keep their eyes off her, whether they approved or not.

Few knew that Tanya had grown so disenchanted with her high school that she had come home from her eleventh-grade classes one day and said, "Dad, I hate school. I want to be done."

Bill Van Cuylenborg labored hard to hide his dismay at the thought of his daughter dropping out a year before graduation. He forced himself to reason rather than rage, to hold his daughter's steady, challenging gaze, with those piercing blue eyes of hers, eyes that mirrored his own. He knew that the angry approach, or a move to impose his authority, would only make her dig her heels in deeper—another way in which father and daughter were all too much alike. He put aside the affectionate humor he usually deployed to cajole his daughter and instead assumed his solicitor's demeanor to cut a deal with her: if she stayed in school long enough to graduate, he'd stop pushing her to go to college. He made it clear that he'd happily finance that option if and when she decided to continue her education, but he'd no longer pressure her to take that path.

"It's a pretty good offer," he said, seemingly relaxed, though his stomach churned.

"It's a deal," Tanya said, and they shook on it. Then she hugged him.

Her yearbook entry the following year read, "Tanya, also known as 'sweetie,' would like to be remembered for her fine sense of sarcasm and many different laughs. After graduation Tanya would like to move away from Victoria and become a photographer. 'Catch ya in the movies, chow babe.'"

She came home in her tuxedo after graduation and told her dad, "I'm done."

"We'll see," Bill said.

It's not that Tanya lacked curiosity, passions, or interests. If anything, she had an overabundance of them, all in competition with one another.

Her love of animals came first, but the allure of photography was nearly as strong. She toted her near-pro-caliber reflex camera and black capsules of 35 millimeter film everywhere. She avidly wrote poetry and frequently paused during routine tasks (or when bored in class) to scrawl scraps of verse in notebooks, provoked by something she saw or heard. But before she decided on anything, she wanted to travel.

Her trip with May to Europe that spring had been life changing. After she graduated from high school, her only definite long-term plan was a return trip, this time for an extended visit to the Netherlands to see her father's birthplace. She hoped to find an au pair gig that included room and board. She wanted to explore the possibility of freelance photography while she was there, too, and dreamed of selling photos to publications back home or in the States. That would be her next step in life, and after that, she told her dad, anything was possible.

"When would you like to go?" he had asked when the subject came up again that fall.

"Soon," Tanya said with a smile, just a few weeks before she and Jay left for Seattle.

MUTUAL FRIENDS HAD introduced Tanya and Jay in July 1987, and they had become nearly inseparable by the time of the Seattle trip. Each had visited the other's home many times and had met their families, though the two sets of parents had yet to meet socially themselves. Tanya and Jay were dating each other exclusively that fall, but Tanya had made it clear she would be heading to Europe on her own before the year was out.

She had confided to May that she really did love Jay, but she was not ready for any sort of long-term commitment. Tanya wanted to go out into the world on her own and discover the right path forward. Then, maybe, she could *consider* something as huge and permanent as marriage. But that was years off, Tanya had told her friend, if ever.

One of May's lasting memories from the weeks before the Seattle trip was of the two of them riding on Tanya's bright red Yamaha Beluga motor scooter, which she prized in part because no one else in their circle thought scooters were cool. May remembers her arms tightly wrapped around her friend's waist as they barreled across a grassy, hilly field, bumping along and going airborne, then crunching back to earth, fishtailing wildly, wind in their hair and their laughter and screams drifting behind them on the breeze.

"We knew we were invincible," May would recall many years later.

"Nothing could ever happen to us. We were always going to be fine. Life was kind of like that then. Tanya always said we should grab ahold of it and make the most of it."

More than thirty years later, May can still recall the smell of the grass and gasoline fumes and ocean salt filling the air as Tanya gunned the scooter's throttle and the little engine belched and grumbled and made them feel they could fly.

On the evening that Jay and Tanya were due to return home from Seattle, May Robson's telephone rang just after dinner, right around the time she expected to hear from her best friend. She grabbed the phone, certain it was Tanya, ready to dish on her brief adventure abroad.

But Tanya did not answer May's happy hello. Bill Van Cuylenborg did.

"Have you heard from her?" Tanya's dad asked without preamble. His familiar deep voice, one she associated with a fatherly warmth and dad jokes, sounded strained. She had never heard him like that before, as if he were lifting something heavy.

"No, n-not a word," May stuttered.

"She was supposed to be home by now, and we haven't heard from her. She hasn't called." Bill paused, then added, "So we're getting a bit squirrelly here."

He didn't want to scare her, May knew. But his mood and tone told her Bill was way past squirrelly and well into terror. Something bad had happened—she sensed it, and so did he. Because Tanya *always* called.

This was the pre-cell-phone era, where constant reachability was not really a thing, but that didn't stop Tanya. Whenever she ran late, Tanya would find a phone—a gas station pay phone, a convenience store, a friendly shop owner—to reassure her parents all was well. In

Europe she called her parents every other day like clockwork. She never chafed at staying in touch and staying close.

May finally managed to say, "What can I do?"

Bill said to tell him right away if she heard anything, but they both knew Tanya would call home first, if she was able to call anyone at all. "If I haven't heard from her by tomorrow," he promised, "I'll go look for her."

As soon as they hung up, May began to cry.

3

Wrong Turn

Jay Roland Cook, with his infectious crooked smile and laid-back demeanor, was Tanya's opposite in many ways: quiet, even-tempered, the one person at a rowdy party most likely to calm a disagreement rather than spark one. In a group setting, he would sit and listen and say hardly a thing for most of the night. But then some inner switch would flip, and he'd take center stage, telling story after story, a stand-up comic seizing the spotlight and shedding his shyness like an overcoat.

The middle Cook child and only boy, Jay had a quiet, studious, focused older sister and a precocious and outspoken younger sister to contend with. Their parents expected all of them to tackle major chores around the house and in the family business, Cook and Talbot, Ltd., Heating Specialists. The Cook kids would take turns assuming responsibility for household chores for an entire week: cleaning the house, washing the vehicles, taking care of pets, cooking the meals, and washing the dishes. Both parents worked, Gordon in the heating business and Lee as a cook at the University of Victoria, and both thought the kids should grow up as their parents had been raised: everyone in the family had to pitch in and contribute. In return, each of the Cook children got an allowance of forty dollars a week, later raised to sixty dollars when they were old enough to begin helping with the family business in addition to housework. Out of that money, they had to buy their own clothes, toiletries, makeup, haircuts, and use what was left for personal spending money. Gordon and Lee would buy big-ticket items—coats and shoes—which would come in the form of Christmas gifts. When they were old enough for high school, each of the Cook children was expected to get a part-time job if they wanted extra spending money. Jay worked throughout high school and continued after graduation with a succession of part-time jobs: delivering pizza, crewing aboard a commercial fishing boat, working for his dad.

His frugal upbringing instilled an appreciation for generosity in Jay—as a beneficiary of others' magnanimity, but also as a benefactor to others. He was quick to treat his friends when they went out, even if it emptied his wallet. And he didn't hesitate to pick up a hitchhiker or aid a stranger with a flat tire or other car problems—a habit that seemed endearing before the trip to Seattle, and ominous after. The more extreme tales of his generosity and devotion to his friends would be recounted many times in the years to come, usually accompanied by laughter, none more than the story of his nighttime pizza delivery from hell.

Among Jay's various part-time jobs was his work at a pizza parlor in Oak Bay. One summer weekend, a disappointed Jay told his friends he'd have to miss a long-planned group getaway to a lake cabin because he had to cover a shift at the pizza place that Saturday night. But as he was about to clock out at ten p.m., he decided he just couldn't miss the party and let down his friends, even if it meant arriving late. So he constructed an extra-large deluxe pizza with everything on it, hopped on his ten-speed bike, and set off for the cabin.

This was not a destination any of his friends would have considered reaching by bicycle, and certainly not late at night. Shortly after Jay departed on this three-hour ride over hilly and dark country roads, the rain began, first a drizzle, then a downpour. With his six-foot, four-inch frame hunched forward against the storm, Jay ended up pedaling with one hand on the handlebars, the other balancing the pizza box on his head, the rain thrumming against this makeshift cardboard umbrella as he squinted and struggled to stay on the road.

At one in the morning, Jay's startled friends heard a pounding at the cabin door. They opened up to find a mud-spattered, bedraggled Jay on the doorstep, teeth chattering beneath long, dark hair plastered to his head. He held out a sagging and sodden pizza box with a damp, ice-cold pie inside.

"Pizza man," he croaked to the stunned faces peering out the door. Then they all laughed.

And yet, when the situation demanded it, this same Jay Cook could also be the most levelheaded and responsible guy in the room. At another party, he noticed that a friend had chugged far too much hard liquor far too quickly. No one else at the raucous gathering noticed anything amiss, but Jay grew alarmed at the way his friend looked and acted. It seemed beyond drunkenness. So Jay had sprinted home, borrowed his dad's truck, then returned and insisted on driving the protesting, staggering teenager to the nearest emergency room. He was unconscious by the time Jay pulled up. The doctors later told the boy's parents that their son had acute alcohol poisoning, and that Jay's quick action likely saved his life.

That was Jay in a nutshell, a mix of the responsible and the irresponsible, keenly observant and oblivious. One day he'd appear at the door to his sister's room dressed in his best clothes and surprise her with an afternoon outing for high tea, seeing to every detail of the extravagant treat. The next day he'd be on an outing with the family and get out of the car when they arrived back home only to discover he had lost his jacket and one shoe somewhere along the way, with no idea how it happened. Only Jay, his family and friends said, could come home wearing just one shoe.

Like Tanya, Jay wasn't sure what he wanted to do with his life long-term. He had spoken often of his desire to become a marine biologist, and his family and friends believed he had the aptitude for it, if not the grades. But Vancouver Island wasn't an easy place to leave, and many never did. Two years had passed since high school, and Jay was still working placeholder jobs. For Jay, his twenty-first birthday, just a month away, served as a signpost. After that milestone, he planned to start figuring out his next act.

JAY LEFT THE house late in the afternoon Wednesday, November 18. He had hastily packed the family's 1977 Ford Club Wagon van, a sturdy, ungainly workhorse that could seat the entire Cook family comfortably plus three more adults and still have room for a furnace in back. It stood out in traffic or in a parking lot not only because of its size but for its unusual color, which most closely resembled a tarnished copper penny. It also sported huge, clunky chrome hubcaps that caught the light.

Jay had thrown two green foam bedrolls in the back, along with sheets, pillows, a blanket, and a comforter—basically all the bedding in his room at the time. The rear area of the van could accommodate two to sleep as easily as it could transport a furnace. Jay also packed road maps, instructions, and directions from his dad, six fifty-dollar Canadian traveler's checks, another couple hundred in cash, and a cashier's check for $758.11 in U.S. dollars made out to Gensco Heating, the furnace supplier in Seattle.

Tanya packed her backpack with a change of clothes, toiletries and makeup, and about sixty dollars in Canadian money. She also brought her Minolta 35 millimeter camera and telephoto lens so she could photo-journal the trip. They'd buy food and drinks and cigarettes on the road, but otherwise they had all they needed for a comfortable overnight round trip.

As the time approached for Jay to leave, only his eighteen-year-old sister, Laura, was home to see him off. She should not have been there either, but she ditched school that day and hid in her bedroom until her parents left. Laura had emerged and just started eating her lunch when Jay strolled into the kitchen. She spent the next five minutes pretending not to hear Jay's comments on the sandwich she had brought home from a restaurant the night before, how delicious it looked and how hungry he was. Laura finally relented and broke off half for her brother.

Between bites they talked about the trip to Seattle. Jay tried to seem casual and relaxed about it. It was just an errand for Dad, he said. Tanya was just coming along to keep him company on the long drive. They'd be back in no time, by dinner the next day. No big deal.

Laura rolled her eyes. Sure, Jay, whatever you say. She knew this was in fact a *very* big deal for her brother, whether he admitted it or not. This was not a little jaunt to one of the other islands, like nearby Salt Spring, but to a big city in another country. Just the two of them.

Laura didn't know Tanya very well, but she had been struck by how smitten her big brother seemed to be. She vividly recalled how she, her boyfriend, Gary Baanstra, and Jay had a conversation five months earlier about two girls he had just met. He had said both seemed interested in going out with him, but he wasn't sure what to do about it. And Gary had asked, "Who can you imagine yourself still wanting to be with five years from now?"

Jay had answered without hesitating: Tanya.

"Then you just answered your own question," Gary said. Jay started dating Tanya within the week.

With their shared sandwich reduced to crumbs, Jay gave his sister a hug and they walked out to the van together. The Cook family had a tradition of bidding an extended farewell to any family member heading out of town: They all would go outside and wave goodbye as the traveler drove off. They would keep waving until their sibling or child or parent was out of sight, making sure to line up far enough out in the street to be visible through the driver's rearview mirror. For Jay's trip, however, it would fall on Laura to uphold the family farewell tradition alone.

She stood by herself in the middle of the street, dark hair flying, golden leaves littering the ground around her sneakers. With big swoops of her arm, she gestured goodbye to her brother, continuing the waves as the van cruised down the long block. As Jay reached the corner and prepared to turn out of sight en route to Tanya's house, Laura saw his left arm come out through the driver's window and wave back. Then the van rounded the corner, and he was gone.

As the days, weeks, months, and years passed, the sense of mystery and loss over what happened next became a central part of Laura's and her entire family's life narrative. The grief also brought Laura and Gary closer, as he became her rock and then her fiancé and, finally, her husband. Leave it to Jay, Laura would later say: her generous big brother had left her one last gift.

OTHER THAN FLYING via seaplane, there were two principal ways for Jay and Tanya to travel from Vancouver Island to Seattle. The simplest and most common route was to take a car ferry to the city of Vancouver on the Canadian mainland, then drive across the border and south on Interstate 5. I-5 runs all the way to Seattle and beyond, right through Oregon, the length of California to San Diego, and then across the Mexican border—the only continuous border-to-border interstate highway in America. The I-5 itinerary would have been a no-brainer: a single-highway journey with no turns to remember or miss, and with an interstate exit just a couple of blocks from their furnace supply house destination.

But I-5 is a big and not terribly scenic highway, with traffic that often bogs down at peak times along the way. So Jay and his dad favored the lower-traffic scenic alternative, though it meant a much more complicated set of directions. First there was the ferry from Victoria to Port Angeles, Washington, the closest U.S. port to Vancouver Island. Then seventy-seven miles of turns, forks, and road changes across four different federal and state highways—and that was just to reach Bremerton. Then came another drive-on ferry with landfall at the downtown Seattle ferry terminal. The last leg was a short but easily confusing drive on surface streets to the furnace supplier on the edge of the city's SoDo—south of downtown—industrial district.

Jay and Tanya's trip began with the easiest and quickest leg: a fifteen-minute drive from Saanich to the ferry terminal in Victoria's Inner Harbour. There they boarded the Black Ball MV *Coho* ferry, named for the prized silver salmon native to the region, bound for Port Angeles. The ferry departed Victoria at four p.m., a half hour before sunset.

The *Coho* is large and often crowded, ferrying up to 115 vehicles and a thousand passengers each trip. Even so, several crew members would later recall seeing the couple on board that day, so there is no doubt Jay and Tanya boarded the *Coho* as planned and made it to Port Angeles.

They docked at five thirty p.m. and breezed through the pre-9/11 U.S. customs process, which involved little more than flashing a driver's license before moving on. Crossing back and forth at the

Canada-U.S. border was so relaxed back then that travelers' names and license plate numbers were rarely recorded.

Jay and Tanya stopped at a market in Port Angeles for some sodas and snacks, then drove east on U.S. Highway 101, which cuts across the northern coastal top of the Olympic Peninsula. The peninsula is a jutting block of land on the country's westernmost edge, shaped roughly like a dog's head tilted up to gaze at the moon. Vancouver Island hovers just above, a hand reaching south to stroke the dog's head, the two separated only by the narrow waters of the Salish Sea. Tanya and Jay's route down the peninsula would take them through some of Washington State's most remote and sparsely populated terrain, thirty-six hundred square miles dominated by forbidding, dense, and dark rainforest, mountainous passes, wild rivers swollen with rain, and the nation's longest undeveloped coastline. Poet Richard Hugo wrote of the region:

> *Rain five days and I love it. . . . Here, the grass explodes and trees*
> *rage black green deep as the distance they rage in. I suppose*
> *all said, this is my soul, the salmon rolling in the strait*
> *and salt air loaded with cream for our breathing.*

Only one town on the peninsula, their starting point in Port Angeles, had a population greater than fifteen thousand at the time, and most had far fewer than a thousand residents. This land remained much as it was a century earlier, a vista of rough and unforgiving beauty, a place that swallows human interlopers far more than it welcomes them.

About forty miles from the ferry port, Jay's planned route called for forking left onto Washington State Route 104, which would have brought them to a floating bridge across the Hood Canal. The Hood Canal separates the Olympic Peninsula from the smaller, adjacent Kitsap Peninsula, where Bremerton, best known for its massive U.S. Navy base and shipyards, was just thirty-six miles to the east. The floating bridge is the only place to cross the canal and the only direct route to the Bremerton ferry that would carry them across the Puget Sound to Seattle. Miss that turn, and travelers have to make a long U-shaped detour down the peninsula and around the southern end of the canal, then drive back up on the Kitsap side.

Jay missed the turn.

He and Tanya continued driving south on the 101 instead, alongside the canal, moving far off course through the lush, rain-soaked forestland of the Olympic Peninsula. After about fifty miles—most of the entire length of the Hood Canal—Jay realized he should have reached the bridge by then. Indeed, he should have reached Bremerton by then.

He and Tanya pulled over sometime between eight and eight thirty p.m. at a roadside convenience store in tiny Hoodsport, Washington, to ask how much farther they had to go to reach the turn to the canal bridge.

"Oh, you're a little past that," Judy Stone, the clerk working the night shift at the Hood Canal Grocery, told Jay and Tanya. "A *long* way past that."

Stone would later say the tall young man seemed more rueful than upset about the mistake and the extra driving it entailed. She told the couple they could drive all the way back up the canal to reach the bridge crossing. Or, she suggested, at this point it might be easier just to continue on south a few more miles and go around the base of the canal, then take the highway north up the Kitsap to Bremerton. Jay thanked her so enthusiastically for helping them get back on course, it made the clerk blush.

Stone immediately took a liking to the polite and talkative Jay, which is why she was able to recall the brief meeting days later when the police began to retrace his and Tanya's route. The night clerk enjoyed chatting up her customers in order to while away the lonely late-shift hours, and Jay was happy to oblige. He explained the purpose of their trip, telling her all about his father's heating business and where they were headed and why. He was an open book, Stone thought. There was no guile in him. While Tanya browsed the aisles for snacks, Stone remarked on Jay's baseball cap, which she thought had an unusual style. And that led Stone to talk about her own son and his baseball cap collection as Jay listened attentively.

All the while, Stone noticed someone else in the store, a man in his twenties or thirties in a brown raincoat. She had never seen him before and he did not look like a local to her, and Hoodsport was the sort of place where everyone knew everyone, particularly if you worked in the town's lone grocery market. He came in right behind Jay and Tanya but didn't stand with them in the aisles. And unlike the Canadian couple, this other man most definitely was not friendly or chatty.

Tanya bought two Pepsis and a bag of Skittles candy. Jay got two bottles of Squirt soda and a bag of Ruffles potato chips. They each paid with Canadian cash and walked out.

Raincoat Man left right after them—without buying anything. Much later, when questioned by the police, Stone could not recall seeing any interaction that suggested Raincoat Man was traveling with or even knew Jay and Tanya. Nor did she see him near the van. Whether he might have been following them, Stone couldn't say one way or the other.

About two minutes after leaving, Tanya ran back into the store and asked for a receipt. They needed to keep track of their expenses on the trip, she explained. The cash register tape dispenser wasn't working that night, so Stone came up with a handwritten receipt for seven dollars for "food and supplies" and handed it to Tanya.

Two customers, Jim Franklin and his adult daughter, Patricia, were in the store browsing for a video rental at the time. Both would later remember Jay and Tanya vividly—Jim, because Judy Stone asked him if he knew the correct exchange rate for Canadian money, and Patricia, who recalled, "I was pissed off," because the big copper-colored van blocked her truck from leaving the store lot while Tanya ran inside for the receipt. Jim Franklin would later remember thinking he might have seen a third person in the back of their van, although he could not be sure because it was dark and the van interior was not clearly visible through the windows. Nor could he tell if the person was wearing a brown raincoat. His daughter couldn't tell if there was a third person in the van or not, but she had a good look through the van's windshield right before it pulled out. Patricia saw Jay driving, looking straight ahead and not saying anything, while Tanya sat in the front passenger seat, talking. Her head was turned to the left and she appeared to be speaking back over her left shoulder as the van drove out of sight.

THE OLD FORD van had upgraded stereo speakers, something Lee Cook insisted upon in all the family vehicles, but radio reception on the wild peninsula could be sketchy. This wasn't just the pre-mobile-phone era. It was also an era without satellite radio, Spotify, Pandora, or iTunes. The first iPod, with its mechanical scroll wheel, was four-teen years in the future. The infant form of the web was two years away, and streaming was something water did, not music. Indeed, in 1987, CDs hadn't yet caught up with vinyl records as a preferred for-mat, though they soon would leap into the lead. So the van's old ana-log tuner and cassette tape player were the state of the listening art then, and the chart toppers getting the most broadcast play that week included the audio caffeine of Billy Idol's "Mony, Mony" and Whitney Houston's "So Emotional," Bill Medley and Jennifer Warnes's buttery "(I've Had) The Time of My Life" (the finale song of the then-new film *Dirty Dancing,* which Tanya and her friend May had just seen together), and "Where the Streets Have No Name," by one of Tanya's favorite bands, U2. They brought their own music along, too, including some home-mix cassette tapes featuring Tanya's favorite singer, Smokey Robinson. Tanya practically wore out her recording of "The Tears of a Clown," which she knew by heart, though it debuted two years before she was born.

She also kept one of her trusty blue-lined spiral notebooks in her backpack to help pass the time. This one was filled with doodles, notes, and observations, hastily scrawled poem fragments (including some lines about elusive, deceptive smiles that may have been an homage to "The Tears of a Clown"), and her ever-present "Things to Do" list.

Tanya was seriously into lists. This latest had a hefty twelve tasks she wanted to accomplish, some during the trip, the rest once she got back from Seattle. The number-one item on her to-do list was a reminder to write a letter to her childhood friend Stephanie in Vancouver, as she had fallen behind in their regular correspondence. She also wanted to write to another friend, Miranda. Other entries on the list showed she wanted to buy a new bag, get her mom a present of an album by Cat Stevens, return a borrowed sweater, call her doctor for an appointment, call her manager at Pickwicks Restaurant, buy new underwear, tights, and a bra, get her beloved Beluga motor scooter fixed, and call the Saanich police about her missing ten-speed bicycle. But whether the tasks were to be completed by shopping in Seattle, writing letters in the van, or making calls after returning home, Tanya never got to check off a single item on her last list.

AFTER LEAVING THE Hood Canal Grocery, the travelers took Judy Stone's advice. They continued south, rounding the end of the canal at the town of Shelton, then turning north again onto the Kitsap Peninsula via Washington State Route 3. They felt they were on the right track, but their confidence in their navigational skills had been shaken. So thirty miles later, at nine p.m., Jay and Tanya pulled into another store, Ben's Deli in the town of Allyn. They asked the clerk, Kara Hopper, if they were on course to Bremerton, and she assured them that they were. Their destination wasn't much farther, Kara told them, and then she went over the correct route with Jay just to make sure they could get there without further complications.

Once again Jay seemed talkative and charming to the clerk on duty at Ben's. He told Hopper about the trip from Canada to get a replacement furnace, how they missed their turn, and how they planned to sleep in the van and make the pickup in the morning. Tanya looked tired, the clerk noticed, as if she had just awakened from a too-brief nap. There were wrinkle-like impressions on her face, perhaps from sleeping on the ribbed surface of the van's vinyl seats or a crumpled pillow. Tanya used the deli's bathroom while Jay pumped the gas.

They bought more snacks and drinks. Jay tried to pay for everything with Canadian traveler's checks, but Hopper was reluctant to accept them. While Jay ran back outside to get cash from the van, Hopper jotted down the license plate number—a safeguard against deadbeats who drive off without paying for the gas. But she needn't have worried. A moment later Jay returned, grinning, and offered two Canadian twenties, which she happily accepted.

The couple appeared neither rushed nor stressed out, the clerk would later recall. Nothing seemed amiss, other than their complaint that the missed turn had led to a much longer and more tiring journey than they had hoped for, adding almost twice as many miles of driving as planned. Hopper didn't see anyone else with the couple, or anyone else visible through the van windows—just two nice kids down from Canada, headed to the big city.

After they drove off, Hopper noticed a folder with three fifty-dollar Canadian traveler's checks inside. Jay had left it on the counter. It was too late for Hopper to catch them by then—the van was miles away. She set the checks aside with a note to her boss and the van's plate number. He'd mail them to the bank in a few days if the Canadian couple didn't come back to claim them first.

Jay and Tanya made the half-hour trip from Allyn to the Bremerton Ferry Terminal without any more stops or missteps, arriving just in time for the last ferry out. They purchased a ticket time-stamped at 10:16 p.m., and the ferry departed nine minutes later. The one-hour crossing would have brought the couple to the Seattle waterfront by 11:30 p.m.

Whether they actually sailed on that ferry or simply bought a ticket they ended up not using remains unknown. They most likely did drive aboard and make the crossing—it's clear the couple and the van ended up on the Seattle side of the Puget Sound at some point. But no witnesses or records can show with certainty that they were on the 10:25 p.m. ferry out of Bremerton.

What is known is that Jay and Tanya did not show up to get the furnace when Gensco Heating opened Thursday morning. Arriving workers saw no copper van. Nor did graveyard shift workers at a warehouse next door see the van or the couple parked there overnight. This doesn't mean they weren't there, just that no one—at least no one who has come forward—can say either way.

Their friends and family had no idea yet, but Jay and Tanya had vanished.

4

Patrol Deputy Scharf

Mid-November 1987
Snohomish County, Washington

Deputy Sheriff James H. "Jimmy" Scharf had been working graveyard-shift patrol for three years when the BOLO on the missing Canadian couple earned a brief mention during roll call: be on the lookout for a 1977 copper-colored Ford Club Wagon van with British Columbia plates. It was just one of many notifications that came over the wires every day, and given that the couple's destination had been Seattle, this one didn't seem terribly relevant to the Snohomish County Sheriff's Office. There was no reason to think otherwise—not yet, at least—and certainly not for a young deputy assigned to patrol sparsely populated farm and forest lands that were on no out-of-towner's must-see list.

After roll call, Scharf checked out his ride, then started his shift patrolling the vast southern half of the county, his rounds taking him past the city of Snohomish, alongside the state fairgrounds, and on down toward Monroe and the King County border, right by the dark arc of High Bridge, then eastward.

These rural graveyard shifts could crush some deputies, but Scharf had always been a night owl, no coffee required; his two-tone department woolens—dark shirt with light tan pocket flaps and khaki trousers with dark piping—never looking disheveled, not even at the end of a shift. He was fit and trim, with a gold sheriff's star pinned proudly to his chest, but his light brown hair had prematurely thinned by his late twenties, and his retreating hairline made him look older and more seasoned than he really was.

This was an asset on graveyard duty. His wife, Laura, who was eleven years his senior, liked it that way, too. The age difference was not readily apparent, and that was fine with Scharf. He had grown up with the music and the friends of two brothers ten and eleven years older than him, and he found he was far more at ease with Laura, and more attracted to her, than he had ever been with women his own age. They had married within a year of meeting.

Scharf's patrol shifts were filled with the routine that all rural deputies in America share—car breakdowns on lonely roads, crashes in rainy or frigid weather, minor thefts and other small-time crimes, requests to back up the small city police departments in his area, and the occasional, dreaded domestic violence calls. He had been in a few life-and-death situations, one of them quite recent, but he had yet to use his big .357 Magnum service revolver other than at the practice range or statewide shooting contests, where he was among the top-ranked in the state. On the job, he preferred to defuse risky situations without drawing his pistol—a record he would maintain across what would be a forty-plus-year career with many arrests of violent criminals but not a single shooting.

Just a year earlier he had been sent on a domestic violence call and ended up locked in a life-and-death struggle with an enormous and unnaturally strong manic-depressive man in a delusional state named Dino Scarsella. Scharf first tried to calm him, only to be beaten and choked into unconsciousness. He recovered sufficiently to summon backup—it took five officers to finally restrain the big man, including an injured Scharf. Throughout what had become an epic siege, he had refused to draw his service pistol in favor of his baton, though this proved ineffective against the big man. He had been told Scarsella was normally a gentle giant when he was on his meds, and Scharf hadn't wanted to use lethal force on a person whose behavior was due to illness rather than intent. (Scharf ended up getting sued anyway, along with the four officers and the entire department, after the man died while in custody at the hospital. He and the others were exonerated in a federal court trial.)

Mostly, though, eight hours of overnight rural patrol in Snohomish County consisted of mile after mile of empty, unlit highway ribboning beneath the rolling box of his Dodge Diplomat police cruiser. His world shrank at such times to the vinyl cocoon of his car interior, lit by the green glow of dashboard lighting, quiet but for the basso purr of the V8, the liquid whisper of tires sizzling over wet asphalt, and the occasional crackle and murmur of the police radio turned down low enough for a library. Beyond his windshield, the world was only visible in the 150-yard tunnels of light painted on the black landscape by the Dodge's twin high beams, as hunched silhouettes of moss-bearded forest trees flashed across his peripheral vision in witness to his passage. This landscape hid its secrets well.

That fall, Deputy Scharf had two overnight shifts a week on south county patrol. He knew every curve and dip on the Woodinville-Duvall Road that brought him there, and he remained especially alert in the area around Monroe and High Bridge, where the access points to the Snoqualmie River were known trouble spots. The state maintained a sprawling prison complex in Monroe that had opened a century earlier as the State Reformatory, then expanded to include the Monroe Honor Farm. There inmates raised and tended dairy cows to provide fresh milk for the prison system and to sell locally. Honor farm convicts were treated as trustees with minimal supervision, a system that sometimes invited abuse. In the dark and isolated areas around High Bridge and other boat landing points, some inmates would creep off to meet with outsiders or girlfriends for drug buys, contraband smuggling, and other criminal activities. It was also a popular drinking spot for kids and gangs, and a place well-suited for the illegal dumping of everything from old engine blocks to toxic waste. Scharf had been instructed to periodically check the place, though the odds of actually catching someone there committing a crime were not great, given that the cruiser's approaching headlights on the dark road made surprise impossible.

That week, all had been quiet in the Monroe area during Scharf's two graveyard patrols. Scharf had seen nothing of coppery vans or missing Canadians or anything else out of the ordinary on those shadowed woodland roads. The only sound as he passed was High Bridge's wooden-plank roadbed, which sang beneath his cruiser as he crossed the river, the *bump-bump-bump* of the boards making the steering wheel vibrate in his hands. He was not on patrol in that area when something did turn up at High Bridge a few days later, something that made the BOLO about the missing Canadians very relevant indeed to the deputies and detectives of Snohomish County.

Scharf would have no involvement with the case, however—not for about twenty years. First he would have to leave patrol for detective work, starting as a crimes against children specialist, then as one of the department's elite major crime and homicide investigators, and then finally as the cold case specialist, when he became responsible for all the old unsolved homicides and missing persons cases in the county, including Tanya and Jay's case. Still, he thought about them in 1987 after the news broke, wondering if he or some other patrol deputy might have passed by that van in the High Bridge area, perhaps parked and cloaked in darkness, and never knew. How easily disaster might have been averted if fate had allowed a random headlight beam or reflection of moonlight to shine on the van that November night. A deputy on patrol, a passing driver, a person peering out their window, or someone leaving for work might have spotted something and lives might have been saved. The idea haunted Scharf.

But he knew so many other factors contributed to what happened, so many seemingly inconsequential events and decisions. They all had to occur just so and in precise sequence, like a once-in-a-lifetime alignment of the planets, without which Tanya and Jay's trip would have concluded uneventfully and there would have been no BOLO, no manhunt, no case at all. First there had to be a broken furnace on Vancouver Island. Then a usually reliable Canadian heating supply company had to fall through, and a Seattle supplier had to have just the right vintage furnace and parts. There had to be a customer who needed that installation before the weekend and a business partner who could not make that happen. Jay had to have just lost a job so he had time to go to Seattle, and Tanya's best friend had to be sick so she could not come and provide strength in numbers. The travelers had to reject a simple, foolproof route in favor of a complicated scenic course where a wrong turn was practically inevitable. Jay and Tanya had to arrive in Bremerton hours late, yet in time for the last ferry to Seattle. Omit or change any one of these links in the chain of events, and the couple from Vancouver Island would never have reached the same spot at the same time as a stranger determined to do evil.

Such is the nature of tragedy, Scharf has learned, built not on a single inevitability or intelligent design but on a mosaic of blameless choice and coincidence, fate assembled blindly piece by piece. Decades after those events, it would become the cold case detective's job to find the missing pieces his predecessors had missed—the hidden clue, the breakthrough evidence, the witness who saw something vital that seemed meaningless at the time but that just might turn mystery into revelation.

BEFORE THAT BOLO about the van went out, most likely the Friday morning after Tanya and Jay's departure, a young construction worker, driver, and auto mechanic by the name of Michael Seat drove toward the Snohomish County line, thirty miles north of downtown Seattle. He cruised along the winding Woodinville-Duvall Road near the town of Monroe, while the daybreak sun shone low through the trees and patches of mist still clung to the road and leafy shoulder. A few miles from High Bridge, something unusual caught Seat's eye.

As he did every workday, he drove by a house that belonged to a close friend's parents, set among the tall, mossy forest growth that lined both sides of the road. But this day, as their driveway came into view, something flared in the morning sunlight that didn't belong, a reflective flash that drew Seat's gaze even as it made him squint and blink. There in the driveway he saw a huge copper-colored window van with large chrome hubcaps. It had been those shiny wheel covers that had caught the sun and his attention.

Seat thought that his friend, who sometimes stayed over at his parents' house or a trailer home in back, must have gotten a new vehicle. His old car was kaput, and he had been bumming rides and using the bus lately. Seat had never seen that distinctive Ford van there before, and he was a self-described "car guy" who noticed such things.

He made a mental note to ask his friend about the van the next time they hung out. Then he returned his attention to his driving and left the big van and the High Bridge area in his rearview mirror.

The van was no longer in the driveway when Seat drove by on the way home that evening. He never did ask about that copper-colored behemoth with the mirror-bright hubcaps, and he never saw it in that driveway again. Seat had forgotten all about it by the time he next caught up with his old pal, Bill Talbott. He never mentioned it to anyone.

Thirty-one years would pass before an old photo in the newspaper would jar the memory loose with a startling and terrible clarity and he would pick up the phone to tell someone at last.

5

Searching

As the evening of Thursday, November 19, 1987, wore on in Saanich with no sign of Jay, Gordon and Lee Cook felt less worried than inconvenienced. Tardiness and Jay, they knew, were old friends.

They talked it over with Laura and their eldest, Kelly, and they all agreed that their affable, generous middle child could also be a bit irresponsible. He eventually got where he was supposed to be, as he had done with that sodden pizza delivery to his friends, but not always when or how anyone expected.

In those first few hours, the Cooks did what came naturally for their family, time, and place: they assumed the best. Jay and Tanya were probably having such a good time on the road and in the big city that they lost track of time. Gordon had even anticipated some sort of delay, which is why he made sure his son had enough pocket money to stay in a hotel for an extra night or two if needed.

If his being late felt perfectly normal, Jay's failure to call home about it fell somewhere on the spectrum between annoying and worrisome.

But Gordon and Lee sought to reassure each other: He really was only a few hours overdue at that point. Jay and Tanya were probably just tired from the long trip, and hunting down a pay phone on dark and unfamiliar terrain likely seemed an unappealing prospect. But Jay would come through in the end, his parents and sisters felt certain, because he knew a family was waiting to get a replacement furnace installed in their unheated house before the weekend. So there was really nothing to fret about.

The Cook family members drifted off to bed by eleven p.m., expecting that they'd wake to find Jay at their sunny kitchen table, where they could all gleefully roast him for being late, then forget the whole thing over a big breakfast.

But when Gordon awoke at one a.m., roused from bed by a gnawing worry, he found Jay's room empty. He padded outside and saw no van in the driveway. The last ferry had docked. Jay, Gordon realized, had not been on it.

He slept fitfully after that, then rose early. "He'll call this morning," Gordon suggested when the rest of the family assembled for breakfast. Or he'll just show up with the furnace and that sheepish, hand-caught-in-the-cookie-jar smile of his, Kelly said. That was Jay, everyone agreed, but now the positive tone felt forced.

FEAR ARRIVED MORE quickly at the Van Cuylenborgs' home that evening. Jean kept checking the time and glancing at the phone. Bill buried his dread with an avalanche of reassurances for his wife. Changed plans and late arrivals were the norm with Tanya, not the exception, he reminded Jean. Besides, Tanya was no vulnerable stranger in Seattle: she knew the city well. Every summer for the last eleven years —since Tanya turned seven—the Van Cuylenborgs had sailed the family boat south from Vancouver Island to Seattle. Tanya dutifully crewed the boat, but it was the shore leave she lived for, the opportunity to shop and explore sprawling Seattle. Plus she was tall and fit, an athletc strengthened and toughened by rigorous basketball and tennis.

"She can take care of herself," he argued. "And so can Jay."

Bill was an accomplished solicitor in Canada's British-modeled legal system, originally in general criminal and civil law, then running a large, more specialized contract and real estate practice in which he represented global corporations, governments, and major airlines. He knew how to build, argue, and spin a case, and he did so that night for his wife's sake, as well as to assuage his own fears. By the time he finished, Bill Van Cuylenborg almost believed it himself.

Jean wasn't buying any of it.

"Tanya would have called," she said simply. The able lawyer had no counterargument for this—there was none. Jean kept voicing variations of this phrase throughout the evening, sometimes barely audibly, a mantra she began repeating without really realizing it. "She would have called. She always calls."

They decided to visit Gordon and Lee Cook, a subdued and awkward first meeting. Bill wanted details about the trip to Seattle, so Gordon outlined the route he and Jay had plotted together. All four parents gamely tried to downplay their fears, but the tells were plain to see—Jean's repeated glances at the Cooks' silent telephone, Lee's wringing of her hands, Gordon's halting speech, Bill's forced heartiness as he listed all the innocent explanations for the kids' overdue return. The conversation drifted to an end as the unspoken dread sucked the oxygen from the room, and the Van Cuylenborgs drove home in silence.

After a dinner the couple barely touched, Bill continued to downplay his own fears to avoid alarming his wife further, even as he discreetly made a series of phone calls, beginning with Tanya's best friend, May. Then he moved on to the rest of his daughter's friends, even Stephanie in Vancouver on the off chance Tanya had called or stopped by there. No one knew anything. Then he reached out to Tanya's twenty-year-old brother, John, who was away at the University of British Columbia in Vancouver. He and his once-pesty little sister had grown closer as their age difference became less meaningful over time, and she had recently stayed with him on campus for several days. But, like her friends, Bill's son had not heard anything from Tanya. He had not even been aware she had gone on the Seattle trip, or that she was dating Jay Cook.

"She would have called you first, Dad, not me," John said. "This isn't like her."

Bill wanted to continue to argue the point, but all he could do was agree. "I'll let you know when we hear from her," he promised his son. "Tomorrow."

But they did not hear from her the next day. Come Friday morning, the telephone remained maddeningly mute in the Van Cuylenborg household. Every now and then Jean would pick up the handset and hold it to her ear, making sure there was a dial tone on the landline, making sure it was working.

Bill, still determined to maintain a semblance of normalcy, put on his suit and tie and kissed his wife goodbye. Then he drove to his law office in Victoria as usual.

Halfway through his first appointment, he realized coming to work had been insanity. He hadn't absorbed a word his client said. With apologies, he hurried through the rest of the meeting, then canceled everything else on his calendar that day and drove home, no longer willing or able to focus on anything but Tanya.

"We're not going to wait any longer," he told his wife, who was not the least bit surprised to see him stride into the house an hour after leaving for a day at the office. "We have to act."

He started by picking up the phone and calling the Saanich Police Department to report Tanya missing.

The response was not what he had expected. The officer who took the call listened politely, then counseled Bill not to worry overmuch. "They're probably just off taking some time together. We've seen it a hundred times. It's too soon to panic."

When Bill pressed his case, explaining the kind of person his daughter was and her penchant for always checking in, the officer said he understood how the Van Cuylenborgs felt. But then he delivered what would soon become a familiar refrain: the police could not do anything until at least three days had passed. It would be different if there appeared to be some evidence of foul play, threats, a car crash, or a medical or mental health concern. None of that applied in this case. Tanya wasn't even a minor who could be classified as a runaway—she legally was an adult and could do as she pleased, which included not calling her dad while she was off with her boyfriend.

The lawyer in Bill Van Cuylenborg understood. The parent in him wanted to rage and scream. He hung up in frustration. Three days! Anything could happen in that amount of time. Three days was an eternity.

He told his wife there was no way he'd sit on his hands for those three days while waiting for the police to accept the Van Cuylenborgs' view that something was terribly wrong. Instead, he would retrace Tanya's route and try to find her himself, and perhaps see if he could get a more forceful response from the authorities on the other side of the border.

He recruited his nephew, Bob DeGoey, to go with him on the search. Jean expressed a fervent desire to join, too, anything to take action rather than sitting at home feeling helpless. But Bill pleaded with his wife to take on what he said would be the hardest job in the days ahead: to stay at home. In those pre-cell-phone times, this was the only way to make sure a call from Tanya would not be missed. If Tanya called with an update or a plea for help or to tell them all was well, someone had to be there to take the call.

"I'll find her," Bill comforted his wife before leaving, though neither had any idea if he could keep that promise.

HE AND HIS nephew left that afternoon, starting their trip just as Jay and Tanya had done: with the four p.m. ferry from Victoria to Port Angeles. Bill and Bob talked to crew members en route and found several who had worked when Tanya and Jay had traveled two days earlier. They recalled the big copper van, and two crewmen recognized Tanya's picture. Yes, the couple had gone that way to Port Angeles on Wednesday afternoon—they were certain. Bill felt heartened by the ease with which he was able to trace their movements on this first leg of their trip.

Bill landed in Port Angeles, hopeful of finding the next clue: a store or gas station or restaurant where someone remembered Tanya and Jay and could point to the next stage of their travels. Instead he found that the vast, sparsely populated Olympic Peninsula was literally the perfect place to disappear without a trace, seemingly designed to confound all attempts to track his daughter's movements. Between towns, Bill saw only dense stands of trees velveted with gray-green coats of moss and fungus. He drove for miles without seeing another car or person. The brooding steel-colored skies of November offered no solace. The kids could have run off the road, broken down, or crashed and be stuck almost anywhere. Bill could drive right by them and not see their stranded van in the thickets of dark trunks and tangled foliage. Or, though he could scarcely believe the police could be right about this, Tanya and Jay could be secluding themselves at any one of dozens of campgrounds or parks, determined to be alone and oblivious to the worry they were causing. Either way, he realized his rescue mission faced long odds unless he could get some assistance.

Once again he looked to the police for hope and help, bulling his way past the front desk at the Port Angeles Police Department and talking his way into a meeting with the chief. This did not go well. Bill and his nephew felt that the chief, in his starched police uniform, was more interested in lecturing than listening. Chief Mike Cleland undoubtedly thought he was being reassuring, but what Bill heard was dismissal and disregard. Beneath the niceties and sympathy, the chief made it clear that he wouldn't commit valuable manpower to track down a couple of young people who, he believed, were most likely out of touch because they wanted to be out of touch. This had happened before, the chief said, and he had been proven right time and again: kids will be kids.

Bill maintained a civil tone, but only barely.

"We're out of here," he leaned over and told his nephew at the first opportunity. "This is a waste of time." Then he said his goodbyes without thanks and fled the office.

But as they were leaving the building, a short, stocky man with a bushy mustache, a tan Stetson, and well-worn cowboy boots approached and said, "I can help you." He turned out to be with the department's search and rescue volunteers, and he did indeed help. He organized a search of logging roads and campsites, with volunteers instructed to look for the distinctive penny-colored van with Canadian plates or anyone who remembered seeing it. The search and rescue director squeezed Bill's shoulder as they parted. He said he was a father, too, and understood all too well what the Van Cuylenborgs were going through. He promised to stay in touch and to call immediately if any leads were uncovered.

Buoyed by the knowledge that someone connected to the police at last took his plight seriously, Bill and his nephew drove down the peninsula on Highway 101, just as Jay and Tanya had done two days earlier. Along the way, they pulled over at a succession of small towns, truck stops, and roadside restaurants to question store owners and waitresses, local cops on patrol, and managers at campgrounds. Bill doggedly sought out anyplace that looked likely to have drawn Tanya and Jay into stopping en route to the ferry at Bremerton. They searched all day and night Friday, slept a few hours in a highway motel, then continued to search all day Saturday. They drove well south of the canal bridge turn to Bremerton, even farther than Jay and Tanya went when they missed their turn—all the way to the port city of Aberdeen, nicknamed the Gateway to the Olympic Peninsula from the southern portions of Washington State. Each time they stopped to ask about Tanya, they were greeted with the same shakes of the head and sad, sympathetic smiles: no one they encountered remembered the Canadian couple or their van.

Bill and his nephew drove right by the Hood Canal Grocery that day, where Jay and Tanya had stopped for directions and snacks Wednesday night. The searchers came within a few miles of the couple's other stop, Ben's Deli in Allyn, as well. It's unclear if Bill visited either place. But if he did, he had the bad luck to stop by when the clerks on duty had not been working on the night Jay and Tanya came in.

On Sunday morning, Bill hired a pilot and plane, thinking that he would cover more ground from the air. They spent hours bumping aloft, trying to spot the van on some side road or camping spot, abandoned perhaps, or stuck on a mountain pass. But the air search led nowhere, and Bill gave up on it.

Throughout his travels, Bill made frequent pay phone stops to check on his wife and to see if Tanya had called home. On Sunday, Jean finally had some news to share: she had heard from the Saanich police. A constable had come over to take a missing persons report at last, as the requisite three days of waiting had passed. The constable interviewed the Cooks, too, then contacted the Port Angeles police to say that an investigation had been opened. This meant Tanya, Jay, and the Cooks' Ford van would be entered into the Washington State law–enforcement computer system as potentially missing. The information couldn't be

placed in the national database yet because the federal criteria for de-
claring someone missing hadn't been met. But getting in the state database was certainly progress, Bill and Jean agreed.

None of this implied that there would be an active search for Jay
and Tanya by authorities in Washington, other than the fruitless
campground search by volunteers from Port Angeles. But if their van
was stopped for a traffic violation or otherwise drew the attention of
police in Washington or Canada, the Saanich and Port Angeles police
would be notified.

Bill returned home to Vancouver Island Monday morning to re-
group. Tanya's brother, John, had come home from college by then to be
with his mother and then to join the search. Gordon and Lee Cook
came over that evening with photos and a detailed description of the
van, including its license plate number.

The next morning, Bill had a stack of posters printed with pictures
of Tanya, Jay, and the van, asking for information on the missing couple. Then he, John, and two nephews took off again, this time driving
down the peninsula to Bremerton and on to Seattle and the couple's
ultimate destination.

They questioned crew members on the Bremerton–Seattle ferry
and in both ferry terminals, as well as at Gensco Heating. No one they
spoke to remembered seeing Tanya, Jay, or the van. They put up the
posters all over downtown Seattle, in coffeeshops and restaurants, on
telephone poles and at bus stops. Only constant movement, the feeling
that something, at least, was being done, allowed Bill to stave off the
thoughts he did not wish to acknowledge, dark imaginings about his
daughter's continued absence and what might have befallen her.

When Bill managed to get in to speak with a detective at the Seattle Police Department, he was relieved to find someone else in law
enforcement willing to treat his daughter's disappearance seriously.
The department took his missing persons report and broadcast a

request-to-locate bulletin to other Washington police departments to raise the case's profile.

Bill left the department with a sense of accomplishment and a bit of hope, eager to phone home and tell Jean that the largest police organization in the region was going to help find their girl.

MISSING
SINCE NOV. 18TH

TANYA
Blond hair
Height 5'10"
Weight 130 lbs.
18 years

JAY
Dark brown hair
Height 6'2"
Weight 165 lbs.
20 years

DRIVING

1977 bronze Ford window van
B.C. license plates
9785 MT

Please contact:

6

Fertile Ground for the Bogeyman

There was good reason for police in the Seattle area to be mindful at the outset when parents reported disappearances such as Tanya's and Jay's. In the 1970s and 1980s (and continuing through the 1990s), Seattle and the Pacific Northwest had become home to an extraordinary number of serial kidnappers, rapists, and killers. Bill Van Cuylenborg's story had, by that point, become a frighteningly familiar and ominous refrain in the Emerald City and environs.

From a practical standpoint, the region served as a predator's ideal habitat: Seattle was a big city with a vibrant, youthful street scene and plenty of potential victims, surrounded by the extensive undeveloped woodlands and wild areas of the Pacific Northwest. Easy escape routes by road and water abounded. It was a region known for its glorious, sun-drenched summer lightness followed by months of perpetual darkness, numbing and endless rains, and crushing, weeks-long periods of sunless gloom. There was no shortage of lonely seasons and isolated places for a killer to avoid the gaze of potential witnesses, stalk victims, and commit crimes in utter seclusion, then dispose of evidence—not to mention corpses—with little chance of immediate discovery.

The smugly smiling serial killer Ted Bundy was the most infamous of this group of predators, with his 1975 arrest and confession to thirty rape-murders around the country (many in Washington) and his chilling friendship with a future true-crime author and Bundy biographer, the late Ann Rule. The writer and killer worked together at a Seattle suicide hotline, and Rule, a former cop, sensed nothing amiss behind Bundy's engaging, sunny, helpful manner.

"If anyone had told me then that I'd been locked up all night alone with possibly the most dangerous man to women in America," Rule later observed, "I would've thought they were crazy."

Many years after Bundy, Ann Rule would once again cross over from writer to witness in another murder case: Jim Scharf's search for Jay and Tanya's killer.

Bundy was far from alone. Indeed, it was one of Bundy's contemporaries, serial killer Gary Addison Taylor, who became the catalyst for reform after one of his twenty suspected victims, a nineteen-year-old newlywed named Vonnie Stuth, disappeared from her Seattle-area home in 1974. Her mother's concerns about foul play were dismissed by the police, squandering a chance to save Vonnie from her abductor and prevent Taylor from killing again. The Seattle authorities vowed to take steps to prevent future cases like Vonnie's in their jurisdiction, but decades passed before those painful lessons took root statewide. The fact that, thirteen years later, only Seattle took Bill Van Cuylenborg's concerns seriously, while authorities in Port Angeles and elsewhere did not, shows just how deep resistance can run to even the most sensible policing reforms.

There were other serial killers active in the region then, some of them even more deadly and prolific than either Bundy or Taylor. Several were suspects in Jay and Tanya's disappearance. There was the

Green River Killer, who preyed on young women in Seattle and Tacoma, abducting them on the highways between the two Washington cities. He would rape and strangle his victims, then dispose of the bodies in forested areas. Between July 1982 and the time of Tanya and Jay's disappearance, he is believed to have killed forty-six times, with no end or clues to his identity in sight.

Operating at the same time, the Coin Shop Killer's known murders totaled eleven men and women from New Mexico to Canada beginning in 1980. Most of his crimes began as robberies of rare-coin dealers, with the proprietors shot to death to eliminate witnesses. But the same unknown killer occasionally broke his pattern, most recently a year before Tanya and Jay disappeared. He kidnapped, robbed, and murdered a California couple bound for the world's fair in Vancouver, British Columbia. The couple was abducted on the same road through the Olympic Peninsula where Jay and Tanya were last seen alive. Their elusive killer used the couple's credit card to shop before vanishing.

And then there was the Spokane Serial Killer, who was well into his sixteen known murders by 1987. Most of his crimes took place in the city of Spokane, in eastern Washington, where he preyed on young, vulnerable women caught up in drug addiction or the sex trade or both. He raped his victims, pulled plastic bags over their faces, then shot them in the head. But, like the Coin Shop Killer, this unknown murderer occasionally broke his Spokane-centric pattern, choosing different types of victims elsewhere in the state, including areas on Jay and Tanya's route from the border.

Given this history, the reaction of the Seattle police should probably have been the norm rather than the exception, even without Vonnie Stuth's cautionary tale. But as Bill Van Cuylenborg discovered, other police chiefs did not feel the same way.

In later years, Jim Scharf would be particularly scathing about his predecessors' practice of waiting three days before considering someone officially missing. It was just a policy, not a legal requirement, and it made no sense to him, not then and not now. The police could have been hunting for that distinctive van Jay and Tanya were driving days earlier, he'd later say, perhaps in time to intervene or, at the very least, to start following a much fresher trail. Bill Van Cuylenborg had done all he could, which was far more than most, Scharf thought with admiration that bordered on awe. But the official failure to act at the pivotal moment meant that, all these long years later, the point where suspect and victims crossed paths still remained a matter of mystery and conjecture that tormented the cold case detective. And finding that intersection point was crucial, for solving the puzzle of *where* would be the first step in answering the ultimate question: *Who?*

In truth, the pivotal moment had passed by the time Bill spoke to the Seattle police. That same day, the first media coverage of the missing couple appeared in a Canadian newspaper, then went out on the U.S. wire services. The report quoted only one law-enforcement source directly, the police chief in Port Angeles, whom Bill had found so exasperating on the first day of his search. In his comments to the press, the chief once again downplayed any concerns that Tanya and Jay had come to harm.

"Far more people are missing because they want to be rather than because something has happened. If there were any evidence of foul play, the vehicle would have shown up. . . . There is no evidence that indicates anything happened except they were doing what they wanted of their own free will."

He would soon regret that callous observation. By the time that article was in print, evidence to the contrary had been found abandoned on a lonely country road an hour's drive south of the Canadian border.

7

The Jane Doe of Parson Creek Road

Vic Wold enjoyed his morning walks along the deserted country roads near his home in the tiny town of Alger, Washington, searching for discarded cans and bottles. Storms, floods, and litterbugs deposited a weekly treasure trove along Skagit County roadsides, embankments, and culvert mouths and in the saturated rainforest undergrowth. By gathering and bagging enough of these disposable containers, Wold could sell his haul to a local recycler. He liked aluminum cans best, recycling-gold in terms of making money, but plastic soda bottles were good, too, and his daily scavenging walks allowed the sixty-six-year-old retiree to bask in a bit of civic virtue while also earning a few extra bucks.

On Tuesday morning, November 24, Wold chose to scavenge along the chill, damp shoulder of Parson Creek Road, a lonely stretch with scant traffic and even fewer houses running between Old Highway 99 and Prairie Road. On both sides of the forest road towering trees stood sentinel, branches dripping in olive-green witch's hair lichen, and

Wold relished the hushed, shadowy wildness of the place. He felt like he was the only man on the planet in country like that, with nothing to fear, nothing to startle or surprise him—at least until that morning.

Parson Creek Road ran atop an embankment along this stretch, high above the forest floor, a ruler-straight length of two-lane blacktop with a freshly painted yellow stripe down the middle, providing a shock of color amidst the muted shades of a gray morning. The embankment sloped down thirty feet or so from the road's shoulder. At the bottom, against a rusted section of old corrugated culvert pipe, Wold saw a ghostly white form stretched out on a patch of dead leaves and brush.

He took a few hesitant steps down to get a closer look, fearing he already knew what it was, then freezing in place when suspicion turned to knowledge. He nearly dropped his plastic trash bag half filled with rattling Rainier Beer cans and muddy soda bottles. Then he scrambled back to the road and speed-walked to the nearest house, where, breathing hard, he asked if he could use the phone to call the sheriff.

"I just found a dead body by the side of the road," he huffed, shock visible on his pale face, and he was let in to make the call.

The first patrol car from the Skagit County Sheriff's Office arrived around eleven a.m. The deputy walked from the road straight down toward the body, getting just close enough to verify that there was in fact a corpse, not an injured person or something else mistaken for a human form. From his vantage point, gender wasn't clear, but the body appeared to the deputy to be a male's. At a distance, the form seemed to be wearing a light jacket and almost impossibly white pants. The deputy clambered back up to the road to his patrol car and radioed for detectives and criminalists to come work the scene.

The overcast weather soon darkened and shifted into a steady, frigid rain as a more experienced deputy, Jim Mowrer, pulled up. He stood at the top of the embankment on the north edge of the tarmac and surveyed the scene. All the trees but the conifers had dropped their foliage, so the dominant ambient sound was the parchment rattle of raindrops on fallen leaves. They covered everything, with even a scattering of leaves atop the body below.

Mowrer looked down to check his footing and spotted two plastic zip ties on the ground about five feet in front of him, on the far edge of the road's gravel shoulder, the type that can be cinched closed but not easily opened. Such ties can be used in the building trades to secure bundles of wires or to tighten joints in sections of ducting—or to serve as temporary, disposable handcuffs. The police used similar sturdier ones all the time. These two on the ground had been joined together, one of the ends pushed through the other's ratchet lock, but the loop wasn't closed—it had been cut, broken, or pulled apart and so no longer formed a closed loop. Was this just a bit of unrelated roadside trash? Or had the body below been bound by this double flex tie, and had it been snapped or pulled apart when the person went down the embankment?

Mowrer looked at the body from above carefully, not wanting to further contaminate the crime scene before the investigators arrived by trudging down there as the first responder had done.

He saw the victim lay faceup, head tilted slightly back, positioned as if the person was cloud gazing. From Mowrer's vantage point, the facial features were obscured by brush. But the hands were clearly visible, and, sure enough, he saw something that bolstered his theory that the flex ties had been used to handcuff the person in life and after.

The hands of the corpse were positioned oddly. They were resting over the groin, but not in a typical hands-folded position. It was the opposite: the palms faced outward. The fingers also curled outward from the body, and the backs of the hands faced each other. Such a position of the hands is strained and uncomfortable for a living person to hold on their own, but it's a familiar posture to anyone who has seen arrestees cuffed behind the back.

"This is an unnatural position," Mowrer wrote in his investigation report the next day, "that would be consistent with hands handcuffed in that position during rigor mortis."

Rigor usually sets in two to four hours after death, depending on the weather, and can last for hours or even several days. If Mowrer was correct, it meant this was no accident, fall, or suicide. It was a body dumped by a killer intent on avoiding detection until he or she was long gone, leaving behind a victim who had been bound and killed well before the body landed next to Parson Creek Road.

The detectives and the Skagit County Sheriff's chief criminal deputy arrived a few minutes later. The lead detective, David Willard, picked his way down the embankment twenty yards distant from the point where the body had apparently rolled downhill, then approached from below the culvert. He wanted to avoid disturbing any footprints or other evidence that might provide a clue as to what happened and who else had been present.

Willard, a crime-scene and forensic specialist at his department, saw immediately that the victim was a young woman, probably a teenager. She appeared to have rolled in a prone position down the embankment and come to rest against the old culvert. The metal of the pipe near the corpse had red streaks that looked like blood. There were no footprints around the body, and none nearby except for the single set left by the first deputy, lending credence to the idea that the killer had dropped her at the roadside and let gravity do the rest rather than carrying the body down to position her by hand.

Willard saw that the victim had not been wearing bright white pants, as it appeared from a distance, but was naked from the waist down. The color of her skin gave that illusion, as gravity caused her blood to settle to the lowest points in contact with the ground. The only clothing below her waist was a pair of heavy gray woolen socks with a vivid red stripe at the calf.

Her legs were straight out in front of her and close together, thighs touching and feet, too, which seemed odd if she had rolled all the way down from the roadside. Had her feet been bound, too, at some point, long enough for rigor to have set in with her legs in that position, so they would not splay or shift even during a fall?

Above the waist the young woman remained clothed. She wore an unzipped light gray fuzzy jacket with black spandex cuffs and, under the jacket, a blue-and-white-plaid flannel shirt unbuttoned at the collar. Under the shirt she had on a white bra, but it had been pushed up over her breasts.

She wore two pairs of earrings, a plain wide-band silver bracelet, a black cloth friendship bracelet, three rings, and a watch on each wrist, a man's and a woman's, each with black leather bands. Everything was silver or pewter. Was the extra watch a boyfriend's? A father's? The back of the watch had no inscription.

A silver chain around her neck held the smallest ring as a pendant, silver and ornamented with a delicate heart instead of a gemstone. It was clearly meant for a child's hand and could no longer fit the victim's littlest finger, so it had been converted to a necklace. The silver heart on the ring was inscribed with the cursive letter "T."

Wiry and agile, Willard did a hands-and-knees search of the embankment from the body up to the road, crawling along what he considered to be the probable path of the body's descent. He looked for bits of evidence—a wallet or purse, articles of clothing, a bullet casing, a discarded drink bottle or crumpled paper that might have the killer's fingerprints on it. He found nothing. He returned to the body and asked the two deputies and the other detective, Gerald Bowers, to come down and help place the victim in a body bag so they could carry her up to the road to the coroner's vehicle that had just pulled up. Bowers, who had photographed the scene from the road shoulder, now took more photographs close-up, forever preserving the violent indifference with which the woman's killer cast aside her remains. Then it was time to prepare the body for departure.

As they did this, one of the deputies checked her jacket pockets and found a deck of playing cards, an eyebrow pencil, and a book of matches, but no identification. There was no wallet or keys, and no purse or pack by the body. No labels on her clothing had a name tag. She was, for the moment at least, a Jane Doe.

When they carefully removed the brush obscuring her face, they saw that one clouded blue eye, the right one, remained open, staring sightlessly up at the officers. Her lips were slightly parted, as if in surprise or midsentence when she died.

They shifted her body and found a small amount of blood under her head in the hard, nonabsorbent clay soil where the body had come to rest. Clearly, obscenely visible was a gaping star-shaped hole in the back of her head, the classic wound from a gun fired at extremely close range or pressed against the skull. She had been executed. Bowers pointed out that there was not nearly enough blood beneath her head for the fatal shot to have been fired where she lay. Nor was there any sign of blood spatter on the road above, although rain could have washed away any such evidence if she had been shot there before rolling down the embankment.

The lead detective and chief deputy had already sketched a plan as they stood on the shoulder of Parson Creek Road in the rain. They would assemble a team of volunteers to return that afternoon to mount a large-scale grid search of the immediate area around the body and the roadside above for physical evidence that might be on the ground or concealed by dirt, brush, or leaves. They would have officers canvass the area to see if there were witnesses who heard or saw anything or anyone unusual in recent days near that spot. The autopsy would be performed, hopefully providing a definitive cause and time of death, the type of weapon used, and perhaps uncover more clues about what had happened. They needed to check missing persons reports, files on known runaways, and other sources of information to figure out who this dead girl had been and where she was from.

Time was of the essence, they knew. In murder cases, if a solid suspect isn't identified within the first two or three days, the likelihood of the case being solved plummets—and those odds aren't that great to begin with in the United States, where four out of ten murderers get away with their crimes and are never caught.

The deputies and detectives placed their Jane Doe into the open body bag, taking care that no possible evidence—hair or fibers belonging to the killer, or soil and leaves that might have come from somewhere else—could fall away before it was found and examined by the coroner's pathologist. Then they zipped the bag closed and carried her up the embankment.

They were experienced cops. All had seen more than their share of the harm that humans inflict on one another, and they were well practiced in steeling themselves against emotional turmoil in such situations. But they knew they had just found someone's daughter, someone's wife or girlfriend, someone's best friend. She had been killed and discarded with no more regard than the old cans and bottles Vic Wold had collected just down the road. And so the gentle care with which these official pallbearers lifted and carried her up that embankment had more than just a desire to preserve evidence behind it.

UNBEKNOWNST TO THE procession bearing Jane Doe up that muddy, treacherous slope, something they very much needed was found around the same time about fourteen miles away in the picturesque town of Bellingham, Washington.

Gaye Taylor, the bookkeeper and bartender at a well-worn dive called Essie's Tavern, helped make the discovery. She had been approached by Old Gus, a regular at the bar who did odd jobs for her from time to time in exchange for beer or a few bucks. Now he wanted to know if Taylor had any work for him that morning.

"I need the lot out back picked up," she answered, by which she meant she wanted Gus to cleanse the bar's lumpy, potholed gravel parking lot of its daily accumulation of trash, cigarette butts, used condoms, and the occasional bits of drug paraphernalia left behind from the night before. Essie's served a sometimes dubious roster of customers in its less than prime location next to the engine noise and diesel fumes of Bellingham's Greyhound bus depot. Gus said yes.

After a few minutes, Taylor went out to the rickety wooden back deck overlooking the parking lot to see if he needed a bucket to hold the refuse. Gus looked up and said, "You might want to take a look at this. It's a wallet, I think. And some keys."

He had found them in the mud, a double ring of keys on a chain, some of them clearly car keys. Nearby he had spotted a stained and soaked cloth wallet with an embroidered design on it—a mandala or something similar, Taylor thought. The wallet seemed to have been out there for a while. She reached down and took the items, then looked inside the wallet. She drew out a small pill container and two plastic cards.

One was a health insurance card and the other was a social insurance card, both of them Canadian. She cleaned them off and tried to clean the empty wallet, too, but it was ruined.

Taylor copied the name and address from one of the cards onto an envelope and slipped the cards inside. Neither she nor Gus knew how many stamps would be needed for a letter to British Columbia, so she couldn't just drop it in a mailbox. Instead, Taylor made a mental note to get to the post office as soon as she got the chance. No doubt, she thought, Tanya Van Cuylenborg would be wanting her cards back as soon as possible.

8

It's Always the Boyfriend

Tuesday evening arrived in Seattle with the usual late-November chill beneath low, dense clouds that masked any hint of moon and stars. The brooding skies brought no rain that night, but they imbued the air with a heaviness, of storms building and biding their time before releasing a torrent. Bill Van Cuylenborg regretted not bringing a better coat, so he felt a bit of relief as he entered the warmth of the ferry terminal with his son, John, and his two nephews. While the others set about hanging posters and talking to concession workers and ferry crew members, Bill walked to a bank of pay phones.

It was after the dinner hour when Bill dropped some coins into the slot and called home. He wanted the latest from his wife, and to pass on the encouraging news that the Seattle police wanted to help. But before he could say more than hello, Jean interrupted, her voice a shaky monotone that chilled Bill all over again, even before the meaning of her words fully sank in.

"They found a young girl in Skagit County, Bill. There's no identification on her. But they think it might be Tanya."

The Saanich constables had come by with the news, saying nothing was definite but the victim matched Tanya's description, Jean told him. Then she half sobbed. "They asked who Tanya's dentist was, Bill. They want to use her teeth."

After Bill had left to search, Jean had sat by the phone, willing it to ring, anxious for news. But then the constables had come, and her waiting turned to dread. This conversation with Bill would set in motion events that might end the awful days of worry and hope, but only by replacing them with something far worse: certainty.

In Seattle, a peculiar clarity settled on Bill then, displacing the restless energy that had possessed him through the endless hours of searching the Olympic Peninsula. Somehow he found the strength to resist dropping the battered black phone from his shaking hand and to tell his wife not to give up, that it might not be Tanya whom the police found, that they could pray it was not her. He did not believe this. He longed for it to be true, and he would indeed pray that it was not his daughter they found, but he was sure it was. He knew it. He felt it.

He asked Jean if anyone was with her, and he was relieved when she said, yes, she was surrounded by friends and family. He told his wife he would find out the truth and talk to her soon.

He fished more coins from his pocket and called the Skagit County detective's number that his wife had supplied. He learned that a body matching Tanya's age and description had been found without identification on a rural road, and that, if it was Tanya, the van and Jay were nowhere to be found. The unspoken implication hovered in the air, but Bill just asked for directions to the morgue. The detective suggested someone else should make the actual identification, that Bill shouldn't see his daughter this way. But Bill was adamant. He would trust no one else with this task. He had to be there to see with his own eyes if this truly was his daughter.

By the time he hung up the phone, the others had gathered around him, sensing from his expression that some terrible turning point had been reached. Bill told them. They walked in silence to the car and began the hour-and-a-half drive to the place in Skagit County where, perhaps, their long search would come to an end.

LATER, WHEN HE was told of the extent of the mobilization in two countries to identify Jane Doe and track down her killer, Bill Van Cuylenborg found more irony than comfort in the recitation of the steps being taken. A law-enforcement apparatus that had been profoundly uninterested in searching for a missing girl while she was still alive had now turned on a dime, sparing no effort or expense on her behalf once they believed she was dead.

Within hours, detectives at the Skagit County Sheriff's Office had led a dozen search and rescue volunteer cadets through a painstaking combing of the Parson Creek Road crime scene. These were high school students, interns, and future department applicants trained to conduct mass searches outdoors for crime-scene evidence. They formed up in a line to cover the terrain in twenty-foot-wide sections radiating from the point where the body was found, a solemn parade formation in sodden timberland. They would each take one step forward, maintaining the line, then examine the brush and ground all around them for any objects that could be evidence. This was a slow process designed to capture everything from the smallest bits of potential evidence to another body secreted in the forest growth and clutter. After everyone had peered at their areas, the line took one step and started another slow scan. At the bottom of the embankment, the line would shift and start the next twenty-foot-wide section. They did this until darkness fell. The detectives would return to resume the search with a second group of cadets the next morning. And the next. And the next. So far their efforts had yielded a moisturizer cream bottle, an old jock strap, and a beer can. They were bagged and tagged, to be examined later for any connection to the murder.

Back at the office, detectives pulled files on violent criminals and sex offenders in the area, as well as similar crimes to see if this killing fit a larger pattern. Even before the autopsy, the victim's missing pants and underwear strongly suggested that sexual assault had been the prelude to murder—and possibly what drove someone to stalk and abduct her in the first place. In the field, the search for potential witnesses around Alger also got under way. And once the story hit the media, tips would begin to pour in. It was an all-hands-on-deck moment for the small sheriff's department—everyone would have to come in and help work the case, or free up others to do so, the chief criminal deputy, Ron Panzero, told his staff.

Panzero went through a stack of police bulletins and missing persons reports and queried the state and national criminal information systems. One case in the mix had stood out to him: a report from a Canadian dad named Van Cuylenborg about his eighteen-year-old daughter, gone by then for six days following what was supposed to have been an overnight trip to Seattle with her boyfriend.

The missing persons report that the anguished dad had pushed reluctant police into accepting had just become the key to identifying the body in the woods—and sparked what would soon be a major international investigation.

Panzero called the Saanich police and, within minutes, felt he was on the right track. Saanich provided details on Tanya Van Cuylenborg's description, the clothes she was wearing the day she left, the purpose of her trip, the van in which she would have been traveling, and information on the missing Jay Cook. Everything fit, right down to a first name that matched the pendant with the letter "T" on it. The

Saanich chief said he'd send two detectives to Skagit with their file and photos to join the investigation, and he'd handle telling the families on Vancouver Island that Tanya might have been found.

Skagit County put out another BOLO on the van and Jay Cook, elevating the search from missing person to murder investigation. There was no benign explanation for Tanya's companion or his van to be in the wind if that was her body beside Parson Creek Road. There were only two explanations for Jay's absence: Either he was dead, abducted, or incapacitated, with a killer at the wheel of the van. Or Jay *was* the killer and he had fled in the van.

The crime scene and execution-style shooting had suggested to investigators a stalker, a stranger, possibly a serial killer rather than an impulsive crime of passion. But it's a truism among homicide detectives that, when a woman in a relationship dies violently, it's usually the husband or boyfriend pulling the trigger or wielding the knife or striking the fatal blow. Killings by strangers make headlines, but that's the exception, accounting for only 16 percent of murdered women in America, according to the Centers for Disease Control and Prevention. Fifty-five percent die at the hands of romantic partners. Jay had to be found, one way or the other.

THE DRIVE SEEMED to go on forever—and perhaps Bill Van Cuylenborg wished it would. Finally, though, he, his son, John, and his nephews pulled up to a wood-and-stone A-frame building fronted by a broad green lawn and a pole bearing an American flag. Their destination looked more like a ski chalet than a funeral home to Bill, adding to the unreality of the moment.

The Jane Doe of Parson Creek Road had been brought to the Lemley Funeral Chapel, a fixture in the Skagit County community of Sedro-Woolley since 1935. Less populous counties such as Skagit often lacked dedicated medical examiner's facilities, and rural counties had a long tradition of leasing mortuary space for use by their coroners. Funeral homes, with their cold storage and embalming stations, could easily be adapted to perform autopsies and store the dead while official investigations were under way.

Tanya's family members filed in, first to the staged hominess of the funeral parlor, where someone explained how the process of viewing the body would go and how everything possible would be done to treat the victim, whomever she might be, with dignity and care. Then Bill was urged once again to let someone less emotionally connected to Tanya make the identification. Bill did not waver. He had to see her. He would not place Tanya in another's hands.

Then he and John walked with their escort to the back, the part of the mortuary the public does not usually see, the part filled with metal surfaces and tile floors with drains. The autopsy had been scheduled for the following morning. To preserve any trace evidence that might identify the killer, the victim remained just as she had been found, right down to the bits of leaves and soil clinging to her neck and face, the tangled hair, the one eye open and staring sightlessly. This was the same blue eye her father had seen staring him down across the tennis court, that frequently crinkled in amusement, and that had looked into his own so many times with such bright promise. Bill didn't know if the old trope about the key events in your life replaying in your mind at the moment of death was true. But he found something much like it happened when you learned your child is dead, a flash of indelible moments large and small, from her birth to her first steps to her crewing his boat to her walking out the door with a wave for the very last time. A boundary was fixed inside him then, one that would dominate the rest of his life: there would be the world with Tanya in it on one side, and the world without her on the other. And he wasn't sure in that moment how he could survive on this side of the line.

With his son standing beside him, Bill Van Cuylenborg nodded slowly and said yes, this is my Tanya. Was my Tanya.

At the time, he thought he just had experienced the hardest thing he would ever endure in life. But he was wrong.

IT WAS LATE in the evening, past ten o'clock, before the Cook family's phone rang with a request from the Saanich police. Could they please come to the station? There was some important information to share.

Lee and Gordon Cook and their daughters, Kelly and Laura, drove together to the department and sat down with two detectives at a table. One of the detectives broke the news that Tanya had been found killed, and that she had been positively identified by her father just hours before. There was no sign of Jay or the van, and an international search was under way.

The family members, though their days had been filled with fear of this moment, just stared back blankly. Much later they would struggle to describe their reaction, the words "shock," "fear," and "confusion" inadequate to convey the violently physical sensation of hope being crushed. The words uttered in that harshly lit room made no sense ringing in their ears. Gordon thought of the Van Cuylenborgs, of Jean, always so welcoming and thoughtful when he came to service their heating system, always offering tea and conversation. How could her daughter, that vibrant young woman they had just seen in their house a week ago, be gone?

The earth had dropped out from beneath their chairs, yet the police still sat there across the table, not unkind, but asking logical, sensible questions, and the family tried to focus. They had to help. Jay was still out there. And so was Tanya's killer. Gradually the questions shifted from the details of the trip and the last time they saw Jay and grew more pointed and probing about his relationship with Tanya. Were they having any issues? Tensions? Arguments?

The Cooks at first just treated this line of inquiry like all the previous questions and answered without a thought. No, no problems, Lee Cook said. Gordon and Kelly just shook their heads. Then Laura spoke up, saying she had noticed them growing closer leading up to the trip to Seattle. She described that conversation in the car with Jay and Gary just a few months earlier, and how she came away with the clear impression that her brother was smitten with Tanya, that he saw in her a possible long-term relationship.

But that only heightened the force of this inquisition. One detective said he had been told that Tanya spoke of having issues with Jay, that she was unhappy, that she even planned on traveling abroad without him. Gordon and Lee Cook exchanged worried glances at this ominous turn and what it implied. Lee was shaking her head now. The police had it wrong, she wanted to say; they were putting a negative spin on the normal byplay of a loving relationship. She sensed where this might be going, but before she could protest, her fears were realized.

"You know Jay will be a suspect now?" one of the police officers asked. "You have to be ready for that."

Three decades later, Lee Cook's voice still shakes at the memory of this moment. After days of wondering and worrying and trying not to fear the worst, of trying to nurture the hope that no harm had come to her boy, this was what they had to face now? Police who had the gall to suggest that, no, we're not worried about something bad happening to your boy, we're worried that he *did* something bad? It was too much. It was too wrong.

"No," Lee said, her voice hoarse. "Whatever happened down there, Jay did not do it. He would have done anything to stop it. But he would not do it."

The family was united in affirming that view to the detectives, one after the other. That is not always the case in such situations, when doubt and defensiveness can show on relatives' faces like splashes of tempera. For what it was worth, the officers in that room could see the Cooks wholeheartedly believed in their Jay. And so they listened patiently to the stories of Jay the peacemaker, Jay the helper, Jay the one who broke up fights. The police nodded, said they understood. It wasn't their case at this point, and perhaps the American detectives would come to agree with the Cooks that Jay was no perpetrator. They hoped it would be so.

"But you still need to be prepared," they said.

Then they warned the Cooks to be prepared for something else: there would likely be a great deal of media attention beginning in the morning. The story of the Canadian sweethearts gone to America, only for one to be murdered and the other missing, would garner intense interest on both sides of the border. Reporters would call, might even show up at the house.

The detectives said there would likely be more questions in the days ahead. The investigators in Washington would probably want more information that only the family could provide.

Gordon said they would help in any way they could.

The family returned home in a state of dread, knowing there was no way Jay had done this to Tanya and no way he would voluntarily cut off communication with his family and friends. But the only other explanation for his disappearance was too difficult to contemplate. They prayed that when the phone rang, Jay's voice would be on the other end, that he had escaped harm somehow, or survived somehow, or had managed to come back to them somehow.

"We've heard absolutely nothing," Gordon Cook told a reporter from *The Vancouver Sun*. "There is an awful tension. We have never felt this before, and we are just worn right out. We just keep hoping."

Then they heard about the van.

9

He's Taunting Us

Wednesday, November 25, 1987
Bellingham, Washington

When Gaye Taylor took a break in the early afternoon, she spread the newspaper on the bar at Essie's. The story about a young woman's body found on a country road fourteen miles away caught her attention, and when she read the victim's distinctive name, Taylor didn't need to check the two Canadian IDs in the envelope she hadn't gotten around to mailing yet. She just picked up the phone and told the Bellingham police that they'd better send someone over right away to get the murdered girl's wallet and keys that had turned up behind her bar.

By midafternoon, cops filled Essie's lot, eager to extract much-needed clues from a second crime scene. The painstaking mass search of the brush along Parson Creek Road had, so far, turned up nothing of value. Maybe Essie's would have some answers.

A crime-scene investigator from the Bellingham Police Department crawled under the dilapidated back deck where the evening patrons at Essie's and the bar next door, Rumors, smoked and drank and

75

tossed refuse over the railing. There, crouched in the dank space below, amid trash, rocks, and scraggly weeds, the investigator found concealed behind a dirty white plastic five-gallon bucket more possessions linked to both Tanya and Jay—and to evidence found near Tanya's body.

There were two plastic flex ties, just like the ones found at the murder scene.

There was a partially full ammunition box containing fourteen .380 silver-tipped Winchester Western cartridges.

There were a pair of latex surgical gloves, used and discarded.

There were three more plastic cards belonging to Tanya in the dirt: her British Columbia driver's license, a 7-Eleven video rental card, and a credit card.

There was a lens cap to a Minolta camera like the one Tanya owned and had brought on the trip, and a black clutch purse with jewelry and cosmetics inside, but no ID.

And there was a piece of a notepaper addressed to Jay, with handwritten instructions from his dad on where to go and whom to contact for the furnace.

The flex ties immediately became a signature piece of evidence that linked the two crime scenes and told investigators that the two plastic ties found near the body were definitely related to the crime and not some randomly dropped pieces of trash by the side of the road. The Saanich police, at Skagit detectives' behest, had found out from Gordon Cook that he did not use plastic zip ties at home or in his business, and none were kept in the van.

This and the gloves persuaded investigators to discount Jay as a likely suspect. Killers wore gloves to avoid leaving their fingerprints and to keep their identity hidden. What good would that do Jay? His fingerprints would be on everything anyway. His identity was known. This was now looking like a stranger killing once more.

When these new bits of evidence had been photographed in place, collected, and bagged, the crime-scene investigator, David Richards, packed them into his patrol car. He would take them to the Bellingham police headquarters to log them into evidence and preserve the chain of custody, that all-important ritual assuring items gathered at a crime scene remained as they were found and could be used in court. Then it all could be turned over to the Skagit County Sheriff's detectives so they could analyze the items for fingerprints and other trace evidence.

But as Richards pulled out of the alley behind Essie's and prepared to turn onto the nearest major cross street, Holly Street, he glanced left and saw something that made him jam on his brakes and gape.

Just minutes before, he had overheard the Skagit detectives talking about the keys that had been found with the wallet. They had sounded excited because two of the keys were for a Ford vehicle, and Tanya and Jay had been driving a big copper-colored Ford van with British Columbia plates. If the killer ditched the keys, he could have ditched the van nearby, too. All this had been news to Richards, as Bellingham was in Whatcom County, not Skagit, and he had not heard much about the case before participating in the search behind Essie's.

And now he was at the wheel of his patrol car staring at just such a copper-colored van with BC plates sitting in a Blue Diamond pay parking lot, barely two hundred feet from the bar. Witnesses would later tell police that the van had been parked there for four days, since Saturday, November 21.

Richards leapt out of his car and looked back down the alley. He could see the cops still standing in Essie's lot and he started waving frantically at them to come down the alley to him. When they looked at him quizzically, he got on his patrol car police radio and barked, "The van! The van! It's right here." And they all came running to see what had been sitting just around the corner all this time.

The van was locked, but the keys that the bartender had given them fit. They steeled themselves to see what awaited inside, fearing Jay—or more likely his body—would be there. But Jay was not in the van. There was, however, a treasure trove of other evidence that began to fill in the blanks of what might have happened to Jay and Tanya after they bought a ticket to Seattle at the ferry terminal in Bremerton.

The disarray detectives saw in the van went well beyond the typical road-trip mess of empty Ruffles bags, soda bottles, Skittles wrappers, and ashtrays stuffed with Camel cigarette butts. The interior looked as if it had been ransacked, with Jay's and Tanya's possessions tossed about carelessly, even violently. At least one bag or backpack had been upended. Among the stuff tossed about on the floor and the seats were a blue jar of Noxzema cream, a roll-on deodorant stick, a toothbrush, an open pack of Macdonald Export A Canadian cigarettes, a stray sock, blue jeans, a pair of brown suede shoes, a pair of white panties, and a small travel bag or purse containing makeup and other possessions of Tanya's, including her passport. There were a half-dozen wrapped tampons scattered around the cargo area, along with one used and bloody tampon tossed aside in the rear of the van. Most ominously, there was a smear of blood on the rear passenger bench seat, and more blood soaked into a striped comforter in the back with the foam bedrolls.

A second pair of panties lay in the back, this one with a zip tie snagged in the fabric. There were four more zip ties nearby, connected together into a single lasso or noose.

Wrapped in the bloody comforter was a pair of women's black knit pants with a stain on one leg. The stain turned out to be semen. Subsequent tests ruled out Jay Cook as the source. If there were any lingering suspicions that the missing Jay might be the killer, this finding ended any doubt. He was a victim, too.

As the van was being searched, word on Tanya's autopsy reached the detectives. The pathologist's first finding held no surprises: the cause of death was a gunshot wound to her head. The gunpowder residue embedded in Tanya's wound and scalp showed that the gun had been pressed against her head, or had been closer than two inches, when the single shot was fired. There was no exit wound, which meant the bullet had rebounded inside her head, causing massive and instantly fatal damage to her brain.

The pathologist said the shooting had been "execution style." The bullet recovered from inside her skull was consistent with the .380 Winchester Western silver-tipped cartridges found under the deck at Essie's. These are hollow-point bullets, designed to inflict maximum damage, because they transform on impact from a pointed projectile that passes through to a flattened hammer that wreaks havoc on living tissue.

No other major injuries turned up in the autopsy. She had not been beaten. There were few or no abrasions on her wrists or ankles from being bound, although such injuries could have been minimized if flex ties were placed over socks or jacket cuffs.

Because of the length of time Tanya's body lay at the foot of the Parson Creek Road embankment, allowing her body temperature to equalize with the environment, a precise time of death could not be determined. The pathologist put it at three to five days before she was found on November 24—which meant she died between Thursday morning and sometime Saturday.

Finally, the pathologist reported some signs of possible sexual assault. There was no bruising or tearing of soft tissue that can be present with violent sexual assault, but he found faint parallel scratches on Tanya's thighs, and semen was found in both vaginal and anal swabs. As with the pants, subsequent tests showed the semen was not from Jay but from an unknown person whom the lab techs labeled simply as "Individual A."

A picture had begun to emerge of a struggle, captivity, rape, and murder in that van. Because the witnesses put the van in the lot in Bellingham on Saturday, detectives surmised that the murder most likely occurred sometime Friday, a day after Jay and Tanya were due to return home, although Thursday and Saturday could not be ruled out as the day Tanya died, either. And the unknown Individual A was the prime suspect regardless.

Investigators found a folder clipped to the driver's sun visor that held the information they needed to trace Jay and Tanya's journey down the Olympic Peninsula to Bremerton then on to Seattle. The folder held purchase receipts from the Hood Canal Grocery and Ben's Deli and a Seattle ferry receipt time-stamped 10:16 p.m., all from the previous Wednesday. Also in the folder were three fifty-dollar Canadian traveler's checks and the $758.11 money order to Gensco for the furnace.

Detectives were dispatched to all those locations to interview deli clerks, store customers, ferry crews, and employees who worked at or near Gensco who might remember Jay, Tanya, or their van and who might have noticed if they appeared stressed or in the company of anyone else. The detectives speculated endlessly. Had the couple picked up a hitchhiker who turned out to be a killer? Offered a ride to a fellow ferry traveler who struck up a conversation with the young couple while crossing the Puget Sound? Had they parked in front of Gensco, then gone looking for a diner or a bar still open in downtown Seattle and encountered a killer in the process? Retracing their travels could lead to information on anyone who was seen with or near the couple, and with luck, one of those interviews would put a face and a name on Individual A.

Meanwhile, the dogged search by volunteers combing the scene of Tanya's murder finally turned up one more piece of physical evidence: a Winchester Western .380 shell casing consistent with the bullet that killed Tanya and with the shells found under the deck at Essie's. An eighteen-year-old Explorer scout and volunteer, Jenny Sheahan-Lee, found it under a leaf nine feet down the embankment from Parson Creek Road.

One thing police did not find in the van, at Essie's, or with Tanya, was any clue as to what happened to Jay. Nor was the weapon used to shoot Tanya found in any of the three crime scenes. The killer had kept the gun—or disposed of it elsewhere.

Blood samples from the van, semen samples from Tanya's body and her pants, vacuum captures of hair and fibers from the van seats and carpeting, and soil samples from the van's tires all were taken for lab analysis using the relatively primitive forensics of the era. DNA typing of the semen was possible at the time, but there would have been little point. There would be no DNA databases to search for years to come. In this era, the police had to catch their suspect first, then see if the crime-scene DNA matched. The same limitation held for the hair, fiber, and soil evidence from the van: police gathered such evidence so that, if they found a suspect, it could be used to help prove the case. Forensic science also allowed bullets to be compared to evidence from other shooting crimes as well as to a murder weapon—if you could find the weapon first.

None of the other forensic tools taken for granted today existed yet in any form. There were no cell-phone-tower records to track the movements of either victims or suspect, no facial recognition software that could be applied to grocery and gas station surveillance cams, no networked digital records of traffic cameras to search in real time. There was no Find My iPhone function that could pinpoint a lost phone anywhere in the world, for there were no iPhones. Nor was there Facebook, Instagram, or anything like them to lurk and comb for clues; back then, social media consisted of online bulletin boards and forums powered by forgotten pre-web services with such names as

CompuServe and Prodigy. AOL wouldn't begin its rise as America's first dominant internet platform for another two years.

The gradual, almost silent increase in police forensic mining of these technologies obscures another profound difference between then and now: in 1987, two years before the Berlin Wall came down, the shrinking of privacy brought about by forensic analysis of these platforms would have sparked near-universal condemnation. Now, they are taken as a given, the price of the technological toys we love in the post-9/11 era.

The reality of 1987 was that criminals had to be identified first with eyewitnesses, informants, or tips from the public, or simply because they were the "usual suspects"—the spouse or romantic partner of the murder victim, or the business partner with a big insurance policy payoff as motive, or some nemesis or rival in work, love, or crime.

Even the gold standard of criminal forensics in 1987, fingerprints, had substantial limitations. There were some rudimentary systems in place at the FBI and some states for computer-aided optically scanned fingerprint searches. Such a system in California had been used in 1985 to try to identify Richard Ramirez, the Night Stalker serial killer of thirteen. But by the time the system generated a list of possible matches that included his name, the San Francisco police had identified Ramirez through witnesses. In the vast majority of cases, the main use for fingerprints, like the rest of forensics, was to confirm suspicions, not to create them.

WITHIN A DAY of its discovery in Bellingham, the Cooks' family van was rolled into a police garage for examination. First it was thoroughly searched, the interior and exterior photographed, the objects within catalogued, and then all the items inside the van, from the cigarette butts crammed in the ashtrays to the coffee cups, empty chip bags, and even the used tampon tossed in the back, were seized, bagged, and tagged. Criminalists, wielding the softest fine-bristled brushes shaped like the small cousin of the old-fashioned feather duster, lightly coated the surfaces capable of retaining fingerprints with a dusting of fine powder—black for light-colored surfaces and white on dark surfaces and glass. The objects from the van that could hold fingerprints were also dusted. Once revealed by the powder, these latent prints were lifted with a piece of transparent tape that was then placed on a white fingerprint card. With some surfaces, or because the traces were too faint, dusting wouldn't work, so another method was used as well: fuming, a then-new process that used cyanoacrylate vapor, better known as superglue, to reveal otherwise invisible prints.

Hundreds of fingerprints were found all over the van, inside and out. None of them proved useful. The killer with his surgical gloves had been careful. No clear prints were found inside the van that didn't belong to Jay or Tanya or someone else in the Cook family. Technicians even tried to recover fingerprints from the inside surface of the latex gloves, but they were defeated there, too, by the killer's use of talcum powder. Ron Panzero would later say he felt the murderer was both arrogant and experienced at covering his tracks—and that he was taunting the police with evidence he left behind knowingly, not carelessly.

"The obvious thing he's telling us is that here, you can have the gloves, you're not going to find any prints," Panzero would later say. "You can have the bullets and shells because you're not gonna find the gun."

But just when detectives felt foiled by the killer, the painstaking examination of the van finally produced a key piece of evidence: investigators lifted not a fingerprint but a single clear palm print from the outside of one of the rear doors to the van's cargo area. Investigators later determined that neither Jay, Tanya, nor any member of the Cook family had left that print.

Assuming it had not been left by some innocent passerby who leaned against the van door while it was parked, the police had in hand a tangible means of tying a suspect to the van. It was an important find.

But the usual catch applied: they needed to identify a suspect first, then see if he matched the mystery palm print. Even the minimal computer databases for fingerprints available at the time couldn't narrow the search: the capability to handle palm prints wouldn't arrive for decades. They needed the killer to slip up, to reveal himself somehow. He might blab to a friend. He might be seen in possession of something of Tanya's or Jay's. Perhaps someone could link him to the van. Finding some such connection became a top priority.

So far, investigators had come up dry on that score. Tips from the public—from ordinary citizens, convicts looking for a break on their own cases, cops in other jurisdictions with similar-seeming open cases—began to pour in after two days of international press coverage. But these tips were uniformly disappointing. Most involved vague and contradictory accounts of sightings of a copper-colored van in various locations around Washington. Several others were from investigators suggesting possible suspects they had their eyes on for other crimes: a state parole agent said that the "word on the street" was that the Canadian couple had been murdered by a violent drug dealer known as "Fat Pat," who specialized in shooting his customers so he could steal their money instead of delivering the drugs. Meanwhile, a detective in Vancouver said he had investigated another shooting involving a criminal who lived in Alger near where Tanya's body was found, and that his suspect might be involved in Tanya's murder. None of these tips led anywhere.

So many tips came in that some received little or no follow-up, even though a few seemed to be credible. One such report came from a worker at a Skagit County potato farm who said he saw a Ford van matching the one on the news pull into the farm's driveway and stop for a minute around nine thirty a.m. on Friday, November 20. The caller said he had a clear view of two people in the van. The driver was not Jay but a man in a dark coat with styled hair who looked thirty-five to forty years of age. The witness said he also saw Tanya—or someone who looked just like her picture—in the passenger seat, her eyes closed, face pale, not moving. She had what looked like a bruise or a birthmark on the side of her face. The van turned around in the driveway and drove off.

The time frame was plausible and the farm was a fifteen-minute drive from Parson Creek Road. But nothing more was ever done with that tip.

Another tip that came in around the same time, however, garnered a great deal of attention and immediate follow-up: a couple who swore they saw both Tanya and Jay at a market in Port Angeles with another passenger in the van. The Port Angeles police conducted the initial interview of the witnesses and told the Skagit County investigators that the account seemed credible. A police sketch artist was dispatched to create a likeness of the mystery passenger in the van based on the witnesses' description, a potentially huge break in the case. But then someone realized there was a fatal flaw in the story: the witnesses said they had seen the people they believed to be Jay and Tanya in Port Angeles on the first day of their trip at ten o'clock in the morning. That would have been seven hours before their ferry had arrived from Canada. It could not have been Jay and Tanya, and the sketch artist turned his car around and went home.

And so Thursday—Thanksgiving Day—arrived with hundreds of tips that provided no breakthroughs, no identification of Individual A, and no clue as to Jay Cook's fate.

. . .

ONE COUNTY SOUTH of Skagit, Scott Walker had plans for Thanksgiving that had nothing to do with crime and mysteries. Eventually he would be going over to his mother's house for the traditional family holiday meal he had looked forward to every year since he was a kid. But the pleasant autumn morning called to him, and he and a friend decided to get in a little hunting first. It was ring-necked pheasant season in Snohomish County, and the twenty-four-year-old school-district maintenance worker wanted to visit his favorite hunting spot near the town of Monroe along the Snoqualmie River, with its tall grass and pleasant hiking.

He and his hunting buddy worked their way south, bagging a few pheasants but still below the limit as the nine o'clock hour approached. They continued on to an area of wetlands and wild grass, a good spot for pheasants marked by a landmark, one that Deputy Jim Scharf knew well from his twice-weekly patrols: High Bridge. The distinctive two-lane wooden plank deck and rough-hewn trestles towered over the grasslands leading up to the Snoqualmie, then marched across the river.

As they rounded an oxbow pond and the bridge came into view, Walker's bird dog, Tess, broke into a run, catching a scent and disappearing into the high grass after slipping through a barbed wire fence. But then the sound of Tess crunching through the dry brush stopped, and she did not immediately return to Walker's side as she typically did. She was a great hunting dog, so he knew something was amiss. He followed her through the brush with his friend trailing behind, emerging close to the wooden piers that supported the bridge.

He saw Tess standing there, staring at a spot close to the tall bridge supports. She appeared uneasy, unsure what to make of what she saw. Sighting Walker, she moved to his side and stayed back, unwilling to approach the form lying in a crushed area of the high grass.

"Is that someone sleeping there?" Walker asked his friend.

That's what it looked like, a person lying comfortably on his side in the grass, a blue blanket partly covering him, faded and torn blue jeans and tan high-tops visible, legs bent at the knees and crossed just above the ankles.

Walker approached the sleeping figure to see if he or she was okay. Then he got a look at one of the person's arms, protruding from the blanket, and saw skin that was ashen gray. No living person had skin that looked like that.

Walker stopped in his tracks and told his friend that they had to go. They had to report a dead person lying beneath High Bridge.

Jay Cook had been found at last.

10

Nowhere Man

The two killings could not have been more different.

Tanya Van Cuylenborg had been dispatched with a single bullet and instant death, with very little overt injury prior to the gunshot. Detectives concluded she had been sexually assaulted, then killed and discarded with cold, impersonal indifference.

Jay Cook, on the other hand, had been battered and brutalized, possibly tortured. Multiple weapons requiring far more strength and effort than pulling a trigger were used. It had not been quick.

Before he knew anything else about the case, before he even knew whose body was lying in the shadow of High Bridge, Snohomish County Sheriff's detective Rick Bart knew this much: this killing was indeed personal.

One of two homicide detectives at the Snohomish County Sheriff's Office at the time, Bart had taken the Thanksgiving morning call. He had apologized to his wife for interrupting their holiday plans, then took the long drive to High Bridge, not knowing as he drove there whether he would find a suicide bridge jumper, an accidental death, or a murder.

When Bart arrived at the scene, seeing a human form curled about thirty feet below the bridge roadbed, it did look as if the victim might have fallen, jumped, or been pushed over the railing. But this was far from certain, Bart knew: while falls from such a height usually inflict serious injury and can kill, most healthy people would survive even greater heights. Statistically, the 50 percent mortality rate for falls doesn't kick in until the distance reaches forty-eight feet.

Bart went down to ground level and began to meticulously document the setting during his approach to the crime scene by shooting multiple rolls of film with a 35 millimeter camera. He moved from the boat access point close by the bridge, followed a muddy unpaved road under the span, then emerged at the spot where the hunters and their dog had first noticed the body. If this turned out to be a murder, he knew he was following the killer's likeliest path.

There was a field of short grass running up to the bridge, interrupted by a five-strand barbed wire fence that Scott Walker's hunting dog, Tess, had passed through earlier. There was more short grass on the other side of the wire, and then, starting at about six feet from the wooden trestles supporting the span, a long strip of tall grass, autumn brown and waving in the breeze. The tops of the stalks ran five to six feet high.

The body lay just over the verge into the tall grass, with stalks crushed beneath, forming an oval gap, almost a nest, with his tan high-tops resting on the short grass. The victim's faded blue jeans were torn at the thighs, revealing his dark blue boxers beneath. His pant legs had been tucked inside his woolen socks, an old hiker trick to foil ticks, mosquitos, and poison ivy.

The light blue blanket that covered the upper half of the body had helped create the illusion of a person napping in the grass—and also covered up the violence done to him. With it moved aside, Bart saw a blood-soaked shirt, battered face, and livid head wounds beneath tangled and matted hair. This young man had died in a ruinous, frenzied attack, not a fall, Bart realized.

While the victim had initially appeared to be lying on his side, without the blanket covering his upper body, Bart could see that his torso was twisted around into an unnatural position so his top half was lying facedown. The wounds on the back of his skull were clearly visible and had been caused by some sort of blunt object.

The medical examiner had arrived, and, after he completed his initial visual examination without disturbing the scene, Bart watched him move the body to better see the face and throat. Wrapped around the victim's neck, leaving vivid red welts ringing his throat, were two red dog collars woven together with thick strands of twine to form a garrote.

Clearly visible through the victim's open mouth and shoved halfway down his throat was a tissue and a pack of Camel Lights cigarettes. Later, when the pack was removed during the autopsy, the cancerous motto on the side, which would be banished from the brand in the next few years, remained legible and, in this context, obscene: "Low Tar Camel Taste."

The pathologist estimated that the time of death could have been as many as five days earlier—around the time of Tanya's death, although there was no way to know for sure at this point. Though the head wounds appeared serious, the autopsy would later reveal the cause of death had been asphyxiation from strangulation and from the cigarette pack blocking the victim's airway. This was, it seemed, a killer perfectly willing to get his hands dirty, who did not flinch at seeing his victim die in his grasp.

That takes an entirely different breed from the common murderer, Bart knew. America leads the world in murder because guns are easy to get and so, so easy to use. They enable impulsive or effortless killing that's over before there's even time to reflect or second-guess or recoil in moral outrage at what is happening, because by the time those thoughts come, someone is already on the ground dead. So it may have been with Tanya.

Strangling someone until they die, however, takes determination. It takes time. It takes a desire that lasts not a fraction of a second before the bang and finish but two or three minutes, sometimes more, of sustained, sweaty, forceful effort. This is nothing like the fifteen-second movie version of a choking death. No audience actually wants to see the prolonged horror of a true manual strangulation. Very, very few people have or want to have the sort of commitment required to actually *do* it.

The killer at High Bridge had it.

Bart searched the victim's pockets. There was no wallet or identification on the body, just a couple dollars in Canadian money and a business card for Gensco Heating in Seattle. These items meant nothing to Bart. High Bridge was more than sixty miles away and in a different jurisdiction from where Tanya's body was found two days earlier, so Bart had no way of linking his Snohomish County murder with Tanya Van Cuylenborg's. The body was, for that first day at least, a John Doe.

But the following morning, detectives in Skagit County heard about the murder victim found at High Bridge and sent someone down to Snohomish with reports on Parson Creek Road and Bellingham. Bart was getting ready to go observe his John Doe's autopsy when the Skagit detective arrived and pulled out a photo of Jay Cook that his parents had provided. One look at the smiling face in that portrait told the veteran investigator that his lanky victim was indeed Jay Cook. From that moment on, the investigation became a joint operation of the two counties, a partnership that would oscillate between close and uneasy for years to come.

Bart's next step was to call the Saanich police to inform them of the discovery at High Bridge and his tentative identification of the body. He asked for dental records and fingerprints to make a definitive determination of the body's identity, but he made it clear this was just to meet legal requirements. He had no doubts about the identity of the young man he had seen crumpled beneath High Bridge.

After the autopsy, he returned to High Bridge. Bart remembered seeing some plastic flex ties near the body. He had attached no significance to them that first day. They looked like mere refuse in a spot where kids in search of a place to party and illegal dumpers averse to landfill fees had trashed the place with everything from empty beer cans to old refrigerators. On his return visit, armed with information about the flex ties found with Tanya, outside Essie's, and in the van, he located and photographed eight of the plastic strips linked together on the ground near where Jay's body had been. Bart found next to them several stones with blood and shreds of scalp and hair stuck to them— the rocks that the killer had used to club Jay.

As much as the killer's methods differed in Tanya's and Jay's murders, two common elements stood out: both bodies were dumped in remote locations where they remained unnoticed for days, and, as in every other crime scene in the case, there were flex ties present, though not actually binding the victims when they were found.

Also of note, Bart thought: The killer chose remote locations, but he did not hide either body. He had to know they would be found. He left them on display.

That evening the Saanich police met once more with the Cooks. They told the family that the identification of the body found at High Bridge was tentative. Later, dental records and fingerprints would make it official: Jay was gone.

"We were holding out hope until the end," Lee Cook told a reporter who called that night. She was numb with shock and had picked up the phone mechanically, without thinking, and merely made audible for the reporter the words that had been playing over and over in her mind. "But I guess it's the end now."

THE ISOLATED HIGH Bridge location, well-known to locals but not on any heavily traveled route, stands less than two miles from Washington State's second largest prison. The sprawling Monroe Correctional Complex housed maximum and medium security inmates in a conventional lockup and minimum security inmates at the honor farm.

The discovery of a murder victim so close to a major prison profoundly influenced the investigation. The idea took root that the killer could be a repeat offender with a substantial prison record and a personal connection to Monroe. This fit with the suspicion that the double murder resembled other recent crimes in Washington, perhaps the handiwork of a serial rapist-killer who traveled with a "murder kit"—flex ties, gloves, bullets, and a gun. Such a suspect would be determined to leave no living witnesses behind, to ensure he would never have to return to the prison.

"This guy is a predator," one detective on the case told a Canadian newspaper. "The way he dropped Jay off near the prison honor farm—this guy is telling us things. There is a very good possibility he has done this before. And he has probably served time before."

Another detective commented that Jay's torturous manner of death mimicked "practices found within prison walls."

There was no hard evidence to support any of this. But in any case lacking witnesses to the crime or solid indicators pointing to a particular suspect, homicide detectives try to develop a theory and a hypothetical suspect to fit the facts and shape the direction of future inquiries. They still pursued the broad range of tips and leads that continued to come in from the public. But the tips and potential suspects that generated the most interest and effort in the case for years to come would be convicts past and present, known serial killers, and possible links to similar unsolved crimes.

Jay's murder in particular—the location of the body, the heightened violence that appeared to some investigators to be driven by rage and ritual—seemed to support this approach. This theory held that, like so many serial slayers, Tanya and Jay's killer likely seemed personable, persuasive, and charming, luring them into a vulnerable situation, revealing his predatory nature only when he had the upper hand and the couple in his power. They could have met on the road or on a ferry or at a restaurant—or in downtown Seattle while looking for a pay phone so Tanya could call home.

The investigators in Snohomish and Skagit did not unanimously embrace all of these hypotheticals. Rick Bart, for one, thought that, instead of a convict, the Monroe location could just as easily indicate a seemingly upstanding local resident familiar with High Bridge from hunting, fishing, or hiking. He also felt there might have been an accomplice involved, because controlling two hostages and the van would be a challenge for a lone assailant. This raised the specter of Tanya being held captive while Jay was killed. Was his death concealed from her, keeping her in an agony of uncertainty and dread while awaiting her own fate? Or was she forced to watch him die, helpless and hopeless?

"That girl went through hell," Bart's Skagit County colleague, Ron Panzero, would later say. "She must be in heaven."

Despite all their theorizing and their efforts, the investigators kept coming up empty. If Raincoat Man or anyone else had been tailing the couple on the Olympic Peninsula or was in the van with them, as witnesses at the Hood Canal Grocery suggested, the police were unable to verify it. The same was true with the ferries, which detectives had ridden nearly a dozen times, interviewing passengers and crew. They distributed photo-illustrated pamphlets in counties, begging witnesses to come forward. A fifty-two-thousand-dollar reward was offered by the Van Cuylenborg and Cook families and their friends, neighbors, and business associates. They put up posters with photos and the reward amount all over Seattle. And still the detectives had no viable suspect, no idea who had killed the sweethearts from Saanich.

He was the nowhere man.

AFTER THIRTY YEARS, Tanya and Jay's killer was still the nowhere man, and Jim Scharf had become the latest in a long line of investigators wrestling with the case. But the suspect the cold case detective sought had not, as his predecessors believed, been motivated by a desire to kill. Scharf sought someone motivated by sex.

As Scharf saw it, everything that happened after the killer spotted the couple—the abduction, the dumping of evidence, the way Jay died, the decision to execute Tanya—flowed from the killer's desire to rape a young woman and get away with it. Scharf had a profile in mind: He sought a suspect seething with resentment and self-imposed isolation. For whatever reason, he had little or no capacity to forge an actual relationship. Whether this was a product of fear, shyness, rage, misogyny, or a personality disorder, Scharf didn't know. What he did read in the evidence was that this killer wanted to abduct a girl, he wanted to force her to submit, and he wanted at least the illusion that the woman he brutalized was a willing participant. And Scharf felt sure the killer acted alone. He would not want to share his victim with anyone.

He had probably been fantasizing and plotting this for years, assembling his "go bag" with the ties, the gloves, the dog collars, the weapon, and the ammunition. Did he troll downtown Seattle looking for runaways? Or perhaps he sought to target a young girl at the ferry terminal?

Or on the ferry itself? Wherever he did his stalking, it brought him into contact with the Canadian couple, and he chose Tanya for his own inscrutable reasons. Did she remind him of someone who had spurned him? Did she say something that resonated with him or angered him? Scharf also wasn't sure if Jay was an unintended complication or part of the plan all along. Either way, Scharf thought that the killer decided the best means of achieving his goal was to tell Tanya her cooperation would save not just herself but Jay, too. For that, she had to believe Jay was alive.

So, after the abduction, Jay was led into the secluded High Bridge area with Tanya left behind in the van, bound with the flex ties. That explained why a man armed with a handgun chose not to use it: Tanya would have heard the gunfire and known Jay was dead, and that she had nothing to lose by trying to fight or flee, Scharf believed.

The evidence at High Bridge explained what happened next. The killer tried to silence Jay by gagging him with whatever was at hand—the pack of Camel Lights, probably Tanya's, who had been known to smoke that brand. But Jay was tall and fit and might have seized this moment to put up a fight, perhaps leading the killer to stumble, panic, and grab something on the ground to subdue his desperate victim. Using a succession of rocks, according to the physical evidence, the killer had stunned the young man, then choked and strangled him. This was not some ritualistic slaying, in Scharf's view, but a poorly executed and improvised ruse. The savagery of the attack on Jay also suggested rage and resentment of someone who got along with women, who talked to them easily, who asked them out without sweating and stammering. He could have encountered the couple on the road at some point—chatted with them, fooled them, lulled them with feigned normalcy and amiability. Perhaps he had observed Jay's easy charm and thought all the while, I'm going to kill you and take your girl, and you can't even see it coming.

Not only did Scharf's theory explain why Jay wasn't shot, it also explained why there were no signs of fighting or defensive injuries on Tanya, whose nature, according to her friends, would have been to fight, kick, scratch, and attempt to escape under any circumstances. But believing there was some hope of them both getting out of this nightmare alive could have been an even more effective restraint than flex ties.

It was hard thinking about that, even for an experienced homicide cop, imagining the fear and doubt and pain this eighteen-year-old may have endured, hurtling across unfamiliar landscapes on dark roads. And he imagined one possible end, when she finally was told she could go free, that she could walk down this lonely road to a town, call the police, and go save Jay. Did the killer get some perverse pleasure from that, seeing the hope mixed with terror on her face as she stepped out of the van onto Parson Creek Road, believing the ordeal just might be over at last, right up until the time she felt the barrel of that gun against the back of her head and the bastard pulled the trigger? Or was he simply being thorough, dispassionately eliminating the only witness, just tidying up loose ends?

The only way Scharf would ever know for sure was if he found the killer. And then he'd have to persuade him to talk.

11

Baby Alpha Beta and the Finder of Lost Souls

November 21, 1987
Anaheim, California

She was in her twenties, homeless, out of money, living in an old car with two young sons, their father in the wind. She was pregnant again. Nine months pregnant. But she didn't show much, and layered against the winter chill in baggy clothes, she knew no one could tell. This lonely secret was both blessing and curse, she knew. It was all too much.

I can't keep this one, the woman had been arguing with herself. There's no money for diapers, or formula, or pretty clothes, or any clothes. I want better than that for this child.

When the time came, she dropped her boys off with a friend and made an excuse for why she'd be gone overnight. The next day, the Saturday before Thanksgiving, she parked in an empty lot and gave birth at five in the morning, cold and alone. She stared down at her new daughter. They cried in tandem.

When the dark sky lightened to gray, she drove the streets of Anaheim, California, looking for signs of life. She found them at an Alpha

Beta supermarket on a busy street ten minutes from the city's great secular mecca, Disneyland. She saw workers bustling around the market, getting ready to open for the new day. This was her chance.

When no one was looking, she placed her two-hour-old baby, wrapped in a faded yellow blanket, by a milk crate next to the store's grimy trash bins. Then she got back in the car and drove off. Someone would find her, the mother figured. Soon.

Sure enough, within minutes the market's janitor followed the sound of soft crying to the trash bins. He stared down at a dark-haired baby on a milk crate, the infant smeared with blood and amniotic fluid, a portion of umbilical cord still attached and draped over the tiny belly. The temperature hovered in the low fifties that morning and the skimpy blanket could not protect the baby's hands, feet, and face. Her fingers and toes were cold to the touch and tinged blue. The janitor shouted for help and other store workers raced over. One tried to warm the small hands in her own, while a coworker ran back inside to call the police. The first officer on the scene put the baby in his squad car and cranked up the heat as high as it would go until the paramedics arrived and whisked the infant to the hospital.

With the excitement ended and the workday resumed, no one at the Alpha Beta noticed an hour later when a woman with an expression both anguished and furtive drove by the store, staring at the trash bins. The mother had returned to make sure her baby had been found. Satisfied, she drove off and never told anyone, not friend nor family, about the child she had left. The father never even knew she had been pregnant.

At the hospital, the neonatologists checked on the baby's condition. Other than a mild case of hypothermia, she was perfectly healthy. The all-too-practiced official response to child abandonment lurched into motion then: the police launched a search for the parents while the county's social services agency sought a temporary foster home.

The next day, the local *Orange County Register* Sunday paper dubbed the child "Baby Alpha Beta."

The baby had come into the world on the same day Tanya Van Cuylenborg most likely left it. And that little girl in her thin yellow blanket would also be the key to figuring out the identity of Tanya and Jay's killer.

A NURSE WHO lived nearby, named Sheri Tovo, read that article and was entranced. The next morning, she offered to foster Baby Alpha Beta until her parents could be located. When all efforts to find the child's birth parents failed, Tovo adopted the girl, and the child's legal name became Kayla Tovo.

Kayla thrived in her adoptive home, happy and well cared for, though once she was old enough to learn the truth of her origins, the facts of her abandonment at first disturbed and angered her. But she eventually concluded there was a reason her birth mother hadn't put her *inside* a dumpster but next to it instead. Kayla realized she hadn't been thrown away. She had been left to be found, left for a better life, left to be saved.

But after growing up, serving an army tour in Afghanistan, weathering grievous wounds and trauma from a bomb blast, then returning home to start her own family, Kayla felt the need to search for her roots. It was time, she decided, to find out who had left her at that Alpha Beta, and why.

In 2014, she connected through social media with CeCe Moore, who had abandoned her acting career to become a pioneer in the new field of genetic genealogy, which turned the quaint old hobby of family-tree building into a science-driven powerhouse that leveraged the immense information inside consumer DNA tests to uncover family secrets. Moore founded the 171,000-member Facebook group DNA Detectives to help people like Kayla. Foundlings, amnesiacs, adoptees, and donor-conceived children were the group's—and Moore's—specialty. She located their biological families. She unlocked their pasts and their secrets. And she lived in the same California county as Kayla.

Moore had Kayla spit into a tube and send it off to the consumer DNA company 23andMe. Results in hand, Moore was able to reverse engineer a family tree for Kayla, putting her in touch with her birth mother in October 2014. And her brother. And other biological relatives. During an emotional meeting with the woman who abandoned her twenty-seven years earlier, the former Baby Alpha Beta offered forgiveness while her birth mother confessed she had yet to forgive herself. Then Kayla invited her newly discovered relatives to her own son's birthday party several months later, with Moore in attendance as well. The saga had come full circle, from the front-page headlines about the discovery of Baby Alpha Beta just before Thanksgiving 1987 to the family reunion twenty-seven years later, once again on the same front page.

CeCe Moore had accomplished a first: she solved a crime solely through genetic genealogy with techniques of her own devising. Not that she discussed it in terms of a criminal investigation. She told the reporters her volunteer work for Kayla was a search for a foundling's identity, a question of reunion for child and birth mother. But child abandonment is, in fact, a crime. And that's how it would have been treated at the outset had Baby Alpha Beta languished too long alone in the cold outside that supermarket. Then it would have been a murder investigation as well—one that would have remained an unsolved cold case until Moore came along.

Even so, the full import of her accomplishment with Baby Alpha Beta wouldn't be clear to anyone for the next few years—not even CeCe Moore herself—until her path crossed with a cold case detective by the name of James Scharf.

12

She Parts Her Wings and Then She's Gone

They buried Tanya Van Cuylenborg on Saturday afternoon, one day after Jay's mother, father, and sisters learned of his fate.

The Van Cuylenborgs' family and friends, more than two hundred, including such luminaries as the attorney general for all of British Columbia, gathered at the University of Victoria Interfaith Chapel to remember Tanya. People spoke of her love of the outdoors, photography, travel, poetry, and her dog, Tessa. They also spoke of her wit, her endearingly contrarian ways, her infectious laugh. Eight of Tanya's friends recited excerpts from her poetry to the gathering, tears streaming down their faces. Her friend May, looking at Bill, Jean, and John Van Cuylenborg, sobbed, "We'll keep her alive in our hearts."

This was the day Bill Van Cuylenborg realized that the hardest moment in his life had not been gazing down at his daughter's body in a Washington funeral home. It was this moment, this saying goodbye to her in church, surrounded by those who loved her, and then driving home and walking from the car into a house filled with memories of

Tanya. She inhabited every room. She was in the volumes on the bookshelves Bill knew she had read. She was in the sound of the floorboards creaking as if she still paced upstairs, fretting over a line of poetry. And she was in the quiet of her bedroom with its familiar, faint scent of the fragrance she liked best. What was that perfume? Bill couldn't remember, and that small thing hit him with waves of fresh misery, the realization that he couldn't ask her.

Within six months, he and his wife would sell the house and move. What else can you do, he later mused, when you still wake every morning and find your wife weeping next to you?

Tanya was buried within sight of the waves at the Ross Bay Cemetery near Oak Bay, her grave marker inscribed with the image of a dove in flight and a quote from one of her own poems:

She parts her wings and then she's gone.

The reporters were there, of course. News coverage of the case, particularly in Canada, grew intense, with every incremental development and speculation in the investigation reported in the Canadian press on a daily basis, often on the front page in Victoria and the rest of the island. In the days following Tanya's funeral, the families were stunned by a spate of stories suggesting that robbery of a few hundred dollars' worth of money and possessions provided the motive for the killing of their loved ones. In addition to the couple's missing cash, the police found Tanya's green canvas backpack, Minolta camera, and telephoto lens were unaccounted for, as was Jay's black nylon ski jacket with red piping. Later that week, the coverage changed course, focused on the emerging police theory that a compulsive serial killer, not a robber, might be the culprit. In the past three years, three other couples traveling rural parts of Washington had been abducted and slain. Media reports trained their attention on the murders of Robert and Dagmar

Linton, the California couple who disappeared while traveling precisely the same route as Jay and Tanya down the Olympic Peninsula in the summer of 1986.

One week after Tanya's funeral, several hundred mourners gathered again at the same university chapel for Jay Cook's memorial service. His friends and family recalled his kindness and generous spirit, the way his escapades always evoked laughter.

"He was strong," his friend Doran Schiller said at the close, "and he'd want us to be that way, too."

Even when they had suspected the worst, the Cooks hadn't discussed memorials or burial, Gordon Cook told his friends. "We've never had to plan a funeral before."

Yet when the time came and they had to choose, his sisters and parents decided what Jay would want without debate or hesitation. After the service, they took the family boat out and spread their son's ashes at sea, at a spot overlooking the Oak Bay beach where he loved to spend his summers—waters that lapped the shoreline of the nearby cemetery where Tanya was buried.

Then the Cooks went home, where they, like the Van Cuylenborgs, struggled with resuming life in a house that seemed empty without their child.

His twenty-first birthday came on December 16. Jay's mother had already bought and wrapped his birthday present before the trip to Seattle, but now it just sat in her closet. She didn't have the heart to touch it. Instead of hearing the familiar sound of Jay pounding up the back stairs to the kitchen so she could hug and kiss him on his birthday, Lee Cook entered the silence of his bedroom alone. She opened a drawer and pulled out one of his sweaters, pressing her face against its soft wool, smelling the familiar scent of her son.

"There is just too much violence," she'd later muse. "We accept it on a day-to-day basis, we just accept—until something like this happens."

Bill Van Cuylenborg expressed similar feelings and spoke publicly of launching an anti-gun campaign. Mostly, though, the two families waited for news from the police in Washington, for answers to what happened and why—and, most of all, for the detectives to tell them who did this thing and how long the killer would be locked away.

There was a flurry of excitement when Tanya's Rokkor-brand telephoto lens turned up at a Portland, Oregon, pawnshop, identified by its serial number. A Skagit County detective raced to Portland to see if the killer, or someone who knew him, had sold the lens. Could the killer have been so stupid? Was this the big break they had been waiting for?

Those brief hopes were shattered when detectives failed in their attempts to trace the lens back to a person who knew something about the murders. The man who pawned it had gotten it from a friend who worked at a camera shop, and he had purchased it from yet another pawnshop along with other Minolta camera gear. He thought it had come from one of two pawnshops in town, but detectives were unable to trace it back any further. Another disappointing dead end.

Then, on Christmas, the letters began arriving.

The first came inside a holiday card addressed to Gordon Cook at the family home, postmarked as mailed from Seattle. It had a Christmas tree pictured on the front.

Dear Mr. Cook:

As someone who instinctively hates ALL Canadians I couldn't pass up the opportunity to kill JAY and TANYA! Furthermore I'll do it AGAIN if another opportunity presents itself. AND you "AIN'T" NEVER going to catch me. And thanks for the money! I laughed as I "wolfed" down the steaks etc. I've eaten and enjoyed since the fateful evening and morning of November 18th + 19th. Sorry it was one of yours, but I've waited to avenge myself on "any" CANADIAN.

Another Christmas card postmarked from Seattle arrived at the Van Cuylenborg home soon after, addressed to Tanya's dad.

Dear Sir:
I intercepted TANYA and JAY late Wednesday evening November 18th, 1987. I did NOT rape TANYA! Her death WAS quick and painless. As for JAY he was made to suffer. He gagged, choked, cried and begged to be spared. But since he was a Canadian he merited absolutely no mercy or consideration. By the way do you really think that I'd be stupid enough to cash traveler's checks linking me directly to murder!

A third Christmas card went to the Royal Canadian Mounted Police detachment in Victoria.

Dearest Assholes:
I finally got even with Canadians! I murdered Jay Cook and Tanya van Cuylenborg, and the trail is getting cold and colder. By the way, do YOU ALWAYS get your MAN—you don't even always get your FAGGOT. And that makes you lower than a FAGGOT.
    MERRY FUCKING CHRISTMAS and a MUCH BETTER New Year (you need it).
    HA! HA! HA! HA!

Bill Van Cuylenborg stood on the front steps and shook when he saw what was in that first card. Alarmed at the expression on his face, his wife had reached for the envelope. But Bill jerked it away, saying there was no point in her reading such trash. He called the police. Lee and Gordon Cook read their Christmas card together and felt more sad than anything, suspecting a cruel hoax but not knowing for certain. They, too, called the police.

Naturally, the letters and their sender became a focus of investigation, with detectives in both countries researching offenders and hate crimes directed against Canadians. They weren't sure what to make of the tone, which vacillated between hateful and jovial, or the odd grammar and capitalization in the missives, but one thing seemed certain: there would be more to come. Detectives warned the stricken family members to steel themselves—and to turn the envelopes over to the police unopened to preserve fingerprints.

The police were right: the letters continued, becoming more rambling and bizarre over time, frequently juvenile in tone as well as rife with bigotry. A series of New Year's cards came postmarked from Los Angeles next, the first one addressed to Gordon Cook again, containing one of the reward posters that the families had put up around Washington, proclaiming himself a psychopath and expressing hope that true-crime author Ann Rule would write about him someday.

An even more vile and explicit card went to Bill Van Cuylenborg that week, leaving him disgusted and angered, as was clearly intended. More letters and cards followed for months, several timed with Mother's Day and Father's Day, with the latter sent to Bill Van Cuylenborg and signed by "TANYA" in the unmistakably clumsy block-letter handwriting of the author of the other letters. It was in these cards that the writer began habitually using the phrase "Hallelujah BLOODY Jesus" and explaining that he was on a mission from God to rid the world of "Canadian Vermin."

"As fornicators and baby murderers Jay and TANYA deserved to die," he wrote.

On the one-year anniversary of Tanya's and Jay's murders, the writer sent a card to the Van Cuylenborgs addressed to "Dearest Parents," with Tanya's name as the person who signed and sent the letter. "This is an important anniversary! TANYA and I want to remind you that WE LOVE YOU both VERY, VERY MUCH. HALLELUJAH BLOODY JESUS!"

After that, the letters abruptly stopped.

Investigators told the families that they believed the letter writer was probably not the killer. Most likely he was someone who enjoyed notoriety, even of the anonymous kind, but had nothing to do with Jay's and Tanya's deaths. There was nothing in the letters that he couldn't have learned from reading newspaper articles about the case, the investigators said. If the taunts contained nonpublicized information that only the police and the killer could know, that would be a different story.

Even so, the letters were fingerprinted and turned over to a forensic psychologist to profile the writer. Nothing useful came out of either effort. The letter writer remained a suspect—he had, after all, confessed in writing—and his identity became just one more question the investigators were unable to answer.

Little by little, with no progress made in identifying the killer, as one year went by, then two, the case went cold. Detectives stopped working on it daily, then weekly. And the families began to give up hope.

IN AN ATTEMPT to resuscitate their stalled investigation, detectives in Snohomish and Skagit Counties agreed in 1989 to participate in an episode of the TV series *Unsolved Mysteries*. This scripted hybrid show was half public service announcement, half tabloid crime report, with its sometimes lurid narration lent a measure of gravitas through the sonorous voiceovers of actor Robert Stack. Then in its second year on the air, *Unsolved Mysteries* already had a track record of reaching a large audience and generating tips from the public—and from potential witnesses.

The show's stock-in-trade laid in dramatizing murders and other mysteries, using actors in the actual settings where the crimes took place, supplemented by interviews with detectives and family members. Each show had multiple segments about different crimes and weird occurrences. And so, sandwiched between a ghost story about the mysterious Marfa Lights of Texas, an unsolved multimillion-dollar platinum heist, and the hunt for a fugitive mobster, the murders of Jay and Tanya got their fifteen-minute slice of the national spotlight.

At the beginning of the segment, actors portraying Tanya and Jay strolled a ferry deck at sunset, arms around each other's waists, the wind in their hair as they gazed over the railing. The narrator sets the stage: "November 18, 1987. Jay Cook and his high school sweetheart, Tanya Van Cuylenborg, took the ferry from Victoria, Canada, to Washington State. Jay was twenty years old. Tanya was seventeen. They were in love. . . . This was to be their first trip together—they planned on a romantic weekend. Sometime during their journey, Jay and Tanya's peaceful vacation turned into a violent nightmare." Ominous mood music swelled in the background as the show cut to a commercial.

For the Cook and Van Cuylenborg families, participating in interviews for the show was at once surreal and painful, as was watching the broadcast, which they were chagrined to see got numerous details wrong. Tanya's age was incorrect: she had been eighteen at her death, not seventeen as the narrator stated. She and Jay were not high school sweethearts; three grades apart, they had not socialized in high school and had met only after Tanya graduated. The couple was said to have planned a "romantic weekend," but they had gone on a weekday business trip, begun on a Wednesday and scheduled to end the following night. Small details, perhaps, Bill complained. But if the goal was to generate tips and find witnesses, wasn't it important to get the details correct?

Unsolved Mysteries also adopted the latest abduction scenario favored by detectives, who had reconsidered their initial theory that Tanya and Jay might have picked up their killer posing as a hitchhiker as they drove down the Olympic Peninsula. Now, they thought Jay and Tanya might have met their killer on their final ferry ride, where they'd be out of the car and mingling with other passengers. Jay and Tanya were depicted on the show as going inside the ferry lounge from the upper deck when it got too chilly out in the open. A smooth-talking, affable stranger in a dark leather jacket approached and chatted them up, wondering if they were headed to Seattle. When they said yes, he very politely said, "I wonder if you could do me a favor. It's pretty cold out there. I only live a few miles from the ferry dock. Can you give me a ride?"

The ersatz Jay and Tanya smiled and nodded. Sure they would.

"They were friendly, they were young, and on their first trip," said Detective Bart, who a decade later would be elected sheriff, create the cold case unit, and give Scharf the job. "And I think they were easily fooled. An easy mark."

Near the end of the segment, the van drove off the ferry, then parked on a dark street an unknown time later. The stranger, no longer looking so affable, stepped out of the driver's door and walked off into the night alone, ominous music swelling again. No violence was depicted during the show, just another detective providing a foreboding voiceover for this closing scene: "I think it's safe to say that by the time they exited the ferry in downtown Seattle, they probably were in the company of the man who killed them."

None of this conjecture was provable, and there was no reason to favor this scenario over any other. But the show delivered its promised payoff: several hundred tips about the case flooded the *Unsolved Mysteries* toll-free hotline.

Running them down took many months. None of them led any-where.

WHEN HE TOOK over the investigation years later, Jim Scharf's review of the case went beyond reading the archived police files on Tanya and Jay. He also pored over the press coverage and recordings of news-casts, including the *Unsolved Mysteries* episode. For him, the most riv-eting aspect wasn't the actual show, with its clumsy reenactments and Voice of God narration, but what immediately followed: a segment of a Seattle local television talk show, with two guests who had just watched the *Unsolved Mysteries* broadcast as a prelude to being interviewed. Gordon Cook and Bill Van Cuylenborg appeared together on *North-west Afternoon*, the two fathers called upon to display their grief as a means of generating the largest response possible from the public.

The *Northwest Afternoon* hosts were kind and sensitive, but the predictable questions that get asked on such shows are about as subtle as a knife thrust, and no less painful. How has this affected your fam-ily? Do you still have hope the police will find the killer? How do you bear up under such tragedy?

Beneath the harsh studio lights and the unblinking cameras ar-rayed around them, the two men flinched in their chairs as if weather-ing body blows. It was almost unbearable to watch for Scharf, though he could not avert his eyes from his jittery, homemade VHS copy of the show.

"I think we have held up pretty well," Gordon offered. "We're do-ing just about what we've always done. We haven't changed."

Bill, on the other hand, made it quite clear that everything in his life and his family's life had changed forever. He felt consumed by guilt.

"I guess I feel most guilty about not caring enough before about the mayhem and murder. I've practiced law for twenty-odd years, but

I have really only concerned myself with the problem of what do we do with people once they go off the straight and narrow. How do we deal with the criminal? I've never looked at the more difficult part, which is how does the criminal become a criminal?"

As they spoke, the contrast in the two men couldn't be greater. There was the heartfelt but halting Gordon, with his longish brown hair encroaching on the collar of his open-necked blue shirt, often speaking with eyes closed as memory replayed behind his eyelids. And next to him was Bill, with his carefully styled silver hair, sporting a perfect tie and suit in a muted black-and-white plaid. His words were polished, eloquent, and lawyerly—until suddenly they weren't, and the same lost expression that Gordon wore crept across Bill's face for a few moments, an eloquent etching of pain, until the lawyer side took control again. They were not friends, these two bereft fathers—they had never met before the killings, and it was clear that their loss had made any kind of closeness between them impossible. Still, purpose had, for this half hour at least, united them.

Gordon was asked, What kind of boy was Jay? The father just stared a moment, as if unprepared for this most obvious of questions. It was clear the problem was that he had too much to say in response, enough for ten programs, maybe, and it overwhelmed him. So instead, Gordon shook his head and quoted someone else, something safe, something that didn't expose the wound so much.

"His uncle described him to the police the day after it was confirmed . . ." Gordon's words trailed off, leaving unspoken the confirmation of Jay's death, and he shifted gears: ". . . until after they found the van. And he described him as a boy with no rough edges. And I really think that fit. I had never heard it used before, and that struck me as a good description of Jay."

When it was his turn, Bill's words came more easily, or so it seemed.

"Well, of course my view would be extremely biased. She was a teenager. The saddest part that I experienced was all the difficulties, the uncertainties that teenagers face today. She had all the usual questions: 'What is our purpose?' 'Where does the world go?' 'Is there any hope for us?' And she dealt with all of those questions seriously but couldn't find answers. And in the last year of her life, the light went on. She was developing a sense of humor. . . . It was only in the last year she started to be able to participate in looking objectively at problems. And then she graduated last spring from high school. She tried waitressing during the summer. She wanted to go on to have some different experiences in retail sales in the fall and was hoping to go to Europe this spring as an au pair girl. . . ."

He trailed off then. He had strayed from safe ground. At last, his voice cracking for the first time, Bill said, "Certainly Tanya to me was, well, she was the love of my life."

Then he described why they had to leave the family home within six months of Tanya's death. "Because we built the house twelve years ago and the part that Tanya had, her bedroom and bathroom, were certainly hers. Just hers. No one else had ever used it. It took us a long time to be able to go in and to straighten it out and to arrange it. And, finally, I couldn't walk in there without crying, which I did for all the months that we still lived there. . . . We had to move. Although all the memories are good, they were too strong."

The hosts turned to Gordon then, who had been listening and nodding, and they asked if the Cook family had faced something similar. "Well, we haven't cleaned up Jay's room. It's still there. We haven't had the strength, I guess, to do it. But no, I haven't felt that we wanted to leave the house. We've been there for ten years. There's a lot of good memories in the house as well."

Another defining moment came near the end. Bill said that the murder of his daughter led him to change his lifelong opposition to capital punishment. "I think it's time we bring the death penalty back to Canada."

And, naturally, the hosts turned to Gordon and asked if he felt the same way, and Jay's dad said, "No, no," speaking softly, just above a whisper, yet somehow with an undercurrent of strength. The cameras had periodically shifted to show glimpses of the studio audience, and they did so in this powerful moment, showing faces looking stricken, sympathetic, a few tearful.

"We've talked about it many times because of our two daughters, and we all feel that whoever did this must have had a terrible upbringing. And I guess we feel sorry for them in a way. We certainly don't want to see them out on the street. But ... I don't know. I'm not for the death penalty. I wasn't before and I'm not now."

And in that moment, it seemed clear that if either of the two men could survive their grief and find a way to enjoy life again, it would be the furnace repairman, not the attorney.

Three decades after that broadcast, Scharf would remain in touch with Gordon Cook, who still waited for answers. But Scharf would never get the chance to meet or bring news of new developments to Willem Van Cuylenborg, the man who had searched so hard and long for the daughter he still believed to be alive, and then for the killer who took the love of his life. Bill would be twenty years in his grave before there would be a break in the case, his ashes laid next to his daughter's in that lovely old seaside cemetery. He never recovered, his family and friends would say, a tormented man until the day he died suddenly at age sixty-two of a stroke, on May 14, 1997.

The intensity he had brought to his life and work had faded after Tanya's death, this man who grew up during World War II in The Hague and moved to Canada as a teenager, scratching out a living farming and pumping gas so he could put himself through college and law school. He worked as a prosecutor, then started his own thriving law practice, a leader in the Victoria legal community right up until November 1987. Then he rolled back his practice, eventually handing off many responsibilities to his son once he had earned his law degree. Bill reserved his intensity for pursuing Tanya's killer, consulting with private detectives, serial killer experts, and psychics. He plastered Seattle and the Olympic Peninsula with new homemade posters, this time offering a "substantial reward" for information leading to the arrest of the killer.

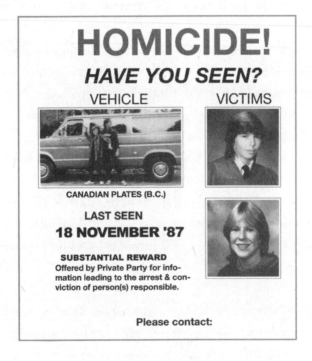

He and his wife, Jean, grew close to Tanya's best friend, May, and maintained a tether to Tanya through her. But then she moved to New Zealand for college and, eventually, marriage, escaping her own years of trauma and sadness with a fresh start, though she returned years later to Vancouver Island to raise her kids. Jean fought with lethargy for many years. "My wife was very outgoing. Now, she doesn't care if she wakes up the next morning," Bill remarked years after the murders. "She finds no purpose in going on in a world where such things can happen. How do you live with that?"

Bill Van Cuylenborg's other great love had been sailing, but even that soured for him. All he could think about when he took the boat out was the annual family trip from Vancouver Island to Seattle, everyone aboard, sunburned and windblown, alive in the salt spray, with the creaking of the masts and the snap of sails going taut as he steered.

"And now? . . . You can only hope to go one day at a time," he said during that Seattle talk show, the last long interview he'd ever sit for. "There really aren't any long-term plans you can make. I can tell you last week was Tan's birthday, and I had to push every hour, one foot in front of the other. Otherwise I would have just sat and felt sorry for myself, which I think you have to be careful of. So all you can do is make the most of the next day, and try to push through."

When Scharf had finished watching the old tape, all the cold case detective could think was how much he would have liked to meet that man, and how he wished he could someday bring him answers.

WHEN THE BIOLOGICAL evidence and samples taken in the stalled case were put into the coldest deep freeze available to Skagit and Snohomish Counties for long-term storage, it seemed to be symbolic of a case gone as cold as it could possibly be.

But the decision to cold-store evidence was, in fact, the opposite of giving up. The care with which this evidence was preserved was in reality an expression of hope by the frustrated forensics experts and pathologists involved in the case at the time. They believed that, in the future, new technology would emerge that could be used to reexamine the stored evidence and, perhaps, solve the case in ways not yet available or even imagined in 1987.

And they were right.

PART II

FINDERS OF LOST SOULS

Why do I do this? It's simple. You see all the lives that were changed by one person who did a terrible thing. You see the devastation, the families haunted for years by questions they can't answer. But the answers are out there. So I work a case until I find them.

—Jim Scharf

13

Cold Case Man

At age twenty, one day after graduating with a two-year degree in applied science and law enforcement, Jim Scharf packed everything he owned into his '68 Mustang. Then he moved out of his parents' home in Granite City, Illinois, and headed west to Everett, Washington, the Snohomish County seat and its biggest city. His older brother had moved there after serving in the army and started a family of his own. He invited Scharf to come out and stay with him until he could get a job and his own place.

Scharf had heard about Washington all his life—his parents had extended family in the Seattle area and had even lived there for a few years before Jimmy was born. And Scharf knew early on that he'd leave his smoggy old iron town for the green, wide-open Evergreen State at the first opportunity.

He found work as a security guard at a local hospital and then learned that the sheriff's department had some openings and was testing applicants. It would mean beginning as a custody officer in the jail, but that's where every rookie started.

He aced the test and got the job, barely beating out a waitress and single mom named Laura, who scored second place and was put on the waiting list for the next hire. Six months later, eager to start policing rather than guarding guys in cells, Scharf left the sheriff's department to take a job as a rookie patrol officer with the city of Snohomish. When he resigned from the jail, Laura got hired for his spot. During the overlap, Scharf met his replacement at a department New Year's Eve party. The two top-of-the-class applicants ended up spending the entire celebration together. Laura was shy and appreciated Jim's effortless conversational skills and obvious admiration of her. They were engaged, married, and settled in Laura's home together that same year.

In 1984, with six years of city policing under his belt, Scharf returned to the sheriff's department as a patrol deputy. Five years later, he landed the job he had wanted all along: detective, assigned to the crimes against children unit. That's where he learned how to be a detective, and it was where he discovered he had a talent for getting suspects to talk.

One of his early cases involved a seven-year-old girl, Ann Marie, who told her school nurse that her dad's boss repeatedly molested her. She said the assaults occurred on family camping trips in the woods, in a cabin, and in a van. In his guise as a family friend, her dad's boss would volunteer to take Ann Marie on hikes and to drive her places when her mother's chronic health problems flared and her father was busy at work. Ann Marie described the where and when of these assaults in remarkable detail for such a young girl: The campground and cabin were in Arlington, she said. The van trip was to Yakima.

A social worker who specialized in interviewing childhood victims of sexual assault had talked to Ann Marie, but Scharf listened in and came away stricken. "We've got to get this guy," he told the social worker afterward. But they both knew how tough it could be to build a solid case with only the testimony of a young child. They had no other witnesses and no clear-cut physical or medical evidence for corroboration. Scharf worried that, if confronted directly, it would be all too easy for the manager to deny everything or lay blame on someone else, most likely Ann Marie's dad. A jury might have trouble deciding what to believe.

So Scharf devised a ruse. First, he waited until after six p.m. to call his suspect and ask him to come to the sheriff's department. The reason: law offices close by then, making it unlikely his suspect could phone a lawyer for advice. Of course, the suspect could just say, I'll need to talk to my attorney first, I'll call you back in the morning, or simply give a flat no. But they almost never did.

Once he had his target on the phone, Scharf told the man, "Your name has come up in an investigation and I need to talk to you, but I can't do it over the phone. I need you to bring your ID to the office so I know I've got the right person, then I can tell you what it's about and you can help me clear this up."

When the man hesitated, unsure what to make of the detective's deliberately ambiguous phrasing, Scharf added, "Look, once you're here and I explain what it's about, if you decide you don't want to talk to me, you can leave. Okay?"

The suspect agreed, and a half hour later they sat in an interview room together on the fourth floor of the Snohomish County Courthouse—sheriff's headquarters. Scharf then spun the man a story: He claimed he was investigating one of the suspect's employees for drug dealing—Ann Marie's father. Scharf said he believed the employee sold narcotics out of a cabin hideout and at a nearby campground in Arlington, as well as in Yakima. (Deception is legally permissible in a police investigation. Undercover cops can't operate without it, and it can be an interrogator's most effective tactic.)

As Scharf had hoped, his suspect believed the story, concluding that he was being interviewed as a possible witness only and that his employee was the detective's target. Better still, the manager leapt to the defense of Ann Marie's father: He's such a great guy! A model employee! He has nothing to do with drugs.

Scharf was thrilled. He listened respectfully, nodding and expressing gratitude that the manager had set him straight about his employee. Meanwhile, the man unwittingly confirmed all the key details Ann Marie had recounted, except, of course, for the actual molestation. Ann Marie's mom was very ill, the manager said, and she could never be far from a hospital. But she loved to camp outdoors. He had a cabin in Arlington that was perfect for camping but just minutes from emergency care, so he took his employee, the wife, and Ann Marie there often, and to a nearby campground as well. And the manager explained that the trip to Yakima had been a business trip in which he had transported Ann Marie in his own van while the dad took care of business. The boss said he could vouch completely for his employee— none of those locations or trips involved drugs or anything remotely suspicious.

At that, Scharf had all he needed. Not only had his true suspect corroborated much of Ann Marie's story, but he had endorsed the dad's good character, and essentially alibied him as well, since in both his account and Ann Marie's, it was the boss, not the dad, who was alone with the girl at the key moments and places. Now it would be hard for the manager to credibly accuse the dad of coaching Ann Marie into lying or committing the assaults himself.

Abandoning the ruse with calculated abruptness, Scharf announced his true intent: "You've been having sexual contact with Ann Marie, haven't you?"

"Well, yeah," the flustered suspect blurted, then stopped himself. "I need to talk to a lawyer."

Sure, Scharf said, you can go talk to your attorney. But I'm keeping your van.

The van was full of incriminating evidence. After some dithering and a visit to his lawyer, the man eventually confessed and was sentenced to twelve years in prison.

Scharf considers this one of his best interrogations, and it helped him perfect his methodology for the future: after-hours requests to meet, misdirection at the outset of the interview, empathy and respect throughout, and then the pounce once the trap was set.

After five years in crimes against children, Scharf transferred to the major crimes unit, which investigated homicides, and his skills at interviewing suspects made him a key part of the team. One of his early homicide cases was a double murder committed on Christmas Eve 1997, the stabbing of a seventeen-year-old high school senior and his father in their home. Scharf had been with his suspect, nineteen-year-old Chad Walton, for more than an hour in the same interview room where he solved Ann Marie's case. Walton steadfastly maintained his innocence, though the detective sensed real regret in him. The younger of the victims had been the suspect's friend.

"I need a cigarette," Walton pleaded between questions.

The man had not been arrested yet and could theoretically cut off the interrogation at any time and leave. Scharf sensed an opening. "Sure, I'll take you to a place where you can smoke."

Once outside the interrogation room, Scharf let the man light up and smoke for a while, then said, "I can see you're upset."

The man nodded, and the detective added, "He was your friend."

Another nod. "You look like you need a hug," the detective said.

And before Walton could say anything, Scharf hugged his suspect. Walton started to cry. When they sat back down in the interrogation room, he told the whole story, admitting to the murders. He thought his friend's father had a large stash of cash in a safe from growing and selling marijuana, which was illegal in Washington at the time. Walton thought that the murders would make him rich. But when he had left them dead and went to the safe, there were no riches. It had all been for nothing.

Scharf had hugged a confession out of a murderer.

Walton later pleaded guilty and was sentenced to life in prison.

"I've seen that motive for murder a lot of times," Scharf would later say. "Guys who convince themselves that somebody has something they don't have."

By the time he left major crimes, he had gotten 125 confessions in a little more than 300 cases.

In 2005, Rick Bart, the former homicide detective turned sheriff, appointed Scharf to the newly formed cold case unit. Scharf was initially one of two, but while his partners would come and go, some promoted to sergeant, others to different assignments, others to new jobs outside the department, Scharf stayed. He had found his calling.

THERE WERE SIXTY-FIVE unsolved cold cases when Scharf started the unit with his first partner, Detective Dave Heitzman. And the very first tip that came in to welcome Scharf to his new job concerned the only double homicide in the cold case files: Tanya Van Cuylenborg and Jay Cook.

A woman called in to say she had been on the same ferry to Seattle as Jay and Tanya the night they disappeared, and she recalled not only seeing the couple but seeing them hanging out with another man. The other day, all these years later, she recognized that same man waiting in a checkout line at her local Walmart store. She was sure of it. She provided a description and gave the time and day she was at Walmart, but she had no name for him or any other clues about his identity.

Scharf sighed when he read the brief report on the phoned-in tip.

While it certainly was possible for someone to remember a face seen briefly two decades after the fact, Scharf knew it wasn't much to go on. But this was the sort of lead a cold case detective had to work with. And sometime even the longest of long shots could pay off.

But not this time. After months spent digging through store surveillance videos and old receipts, identifying the ATM card the mystery man had paid with, and persuading the bank to contact the cardholder, he finally had a name and a phone number. When Scharf caught up with him, the man in the Walmart said he had never ridden that ferry at any time. And after rooting through his records, he offered proof he was not even living in the state in November 1987. The detective had cleared his very first suspect in Jay's and Tanya's murders. While he never thought the Walmart man would prove to be the killer, he had already decided he would pursue every tip, no matter how iffy, because someday the right one would come along, and it might well appear just as dubious as the first.

In 2008, Scharf heard about a way he might improve the quality of tips coming in on cold cases: He could create a deck of playing cards. These cards would be perfectly good for rummy or poker or any other game, while also serving another purpose. Each card, from ace of spades to two of hearts, would contain the story of a different cold case and a photo of the victim. He would choose the fifty most solvable of the cold cases for starters. The decks would be given away for free to the people who knew murderers, their boasts, and their secrets better than anyone else: prison and jail inmates. Prisoners play cards incessantly to pass the time, gambling for cigarettes and contraband. And they are always looking for a way out or, short of that, a way to improve their circumstances inside. Scharf's cards would offer them that possibility.

The ace of clubs was Patti Berry, who left work at Honey's strip club early on the morning of July 30, 1995, and never made it home. Her

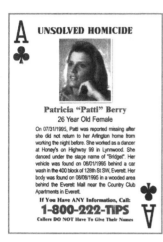

UNSOLVED HOMICIDE

Patricia "Patti" Berry
26 Year Old Female

On 07/31/1995, Patti was reported missing after she did not return to her Arlington home from working the night before. She worked as a dancer at Honey's on Highway 99 in Lynnwood. She danced under the stage name of "Bridget". Her vehicle was found on 08/01/1995 behind a car wash in the 400 block of 128th St SW, Everett. Her body was found on 08/08/1995 in a wooded area behind the Everett Mall near the Country Club Apartments in Everett.

If You Have ANY Information, Call:
1-800-222-TIPS
Callers DO NOT Have To Give Their Names

UNSOLVED HOMICIDE

Susan Schwarz
26 Year Old Female

At about 4:00pm on 10/22/1979, Susan was found dead in her home in the 22800 block of Locust Way in Alderwood Manor. Property was missing from the home.

If You Have ANY Information, Call:
1-800-222-TIPS
Callers DO NOT Have To Give Their Names

UNSOLVED HOMICIDE

**Jay Cook &
Tanya Van Cuylenburg**
20 Year Old Male &
18 Year Old Female

On 11/26/1987, Jay Cook's body was found under the High Bridge near the Monroe Prison's Honor Farm. The body of his companion, Tanya Van Cuylenburg was found on 11/24/1987 in rural Skagit County, south of Alger. Their van was found in a parking lot in downtown Bellingham. Both were Canadian citizens in Washington on business. A ferry ticket was found in the van, indicating the couple may have been on a Bremerton to Seattle ferry on the evening of 11/18/1987.

If You Have ANY Information, Call:
1-800-222-TIPS
Callers DO NOT Have To Give Their Names

blue Honda sedan, splashed with blood inside and out, was found abandoned the next day. A week later, her body was found in a wooded area behind a shopping mall. She had been stabbed many times. Investigators were stumped.

Susan Schwarz was the queen of hearts. Someone entered her home in October 1979 while she showered, tied her up, then fired two bullets into her head point-blank. She was twenty-six. Investigators suspected a burglary gone awry, but they could never find a suspect.

Tanya Van Cuylenborg and Jay Cook were the king of hearts, the only pair among the fifty-two cards Jim Scharf designed.

Each card displayed the sheriff's 1-800 phone number tip line. Each card was an unspoken promise from Scharf: you help me, I help you.

The idea had been inspired by a deck of cards designed by the U.S. Defense Intelligence Agency. Theirs didn't feature unsolved crimes; instead, they had pictures of the most-wanted members of Saddam Hussein's regime during the Iraq War. A detective in Florida adapted the military's idea to his state's cold case unit, gave the cards out in prisons and jails, and quickly solved three open murders. When Scharf read about this, he made sure his department was the next to adopt the strategy. The sheriff's budget had no money for custom-made playing cards, but the detective sought and won a grant from the local Stillaguamish Tribe—enough for five thousand decks in 2008. The reach of the cards expanded beyond the incarcerated to the general public when the Everett-based local newspaper, *The Daily Herald*, featured a different card in the deck, along with details of that case, in every Sunday paper for a year.

And information did flow in, from the public and from inmates.

The first case cracked through the cards was the queen of hearts, Schwarz, whose murder turned out to be a revenge killing that had nothing to do with burglary. A convicted killer who saw the cards in prison called the hotline and turned in his own half brother, Gregory Johnson. An abusive husband, Johnson had been angry at Schwarz for befriending and sheltering his ex-wife, helping her leave town. Once he had a name, Scharf was able to locate a witness whose testimony made a solid case against Johnson, who ended up in prison for life.

The cards helped solve the Berry case, too. Scharf was able to use them as a prop to extract incriminating statements from a repeat offender, Danny Giles, who would be convicted and sentenced to forty-seven years in prison.

Other cards also produced fresh leads. One woman who was missing and presumed dead saw her image on a card and came out of hiding, reconnecting with her family. Scharf had high hopes that the king of hearts and subsequent press coverage could lead to answers in Jay and Tanya's case. Two years after the decks started circulating, the breakthrough finally came. It was August 19, 2010, more than two decades after the murders: Tanya Van Cuylenborg would have been forty-one, Jay, forty-three.

The next day, Scharf would find out just how far this new lead would take him.

. . .

JIM SCHARF'S DAY always begins around six thirty in the morning. His first task—once he exchanges blankety warmth for the morning chill of foothill country—is to grab his thyroid medicine and the full bottle of water he must drink with every dose. It takes a while to get it all down, so he'll multitask by reading the local news online as he sips. Sometimes he dozes a bit in front of the computer. He used to rouse himself with a hearty breakfast, but that simple pleasure is gone. His thyroid meds demand an empty stomach.

Ask him to walk you through his morning routine and he will do so happily, but be prepared for a man who loves a good tangent. He'll bounce between the art of coaxing a murderer to confess ("It's nothing like on TV!"), where to get the best cherry pie, his fascination with the vintage cars he displays in a surprisingly comfortable house-sized warehouse he calls his Man Cave ("It would be neat to own a car from the Brass Era, 1915 or older!"), and the joys of life with pugs. Then there's the tale of his thyroid.

"They took my thyroid out," Scharf likes to say. "Twice."

Seven years ago, he felt a lump in his neck. When he got it checked, the doctors determined he had a cancer the size of a pea and wanted to remove half his thyroid, do a quick biopsy, and if the cancer was confined to the pea, he could keep the other half of the thyroid and he'd get by just fine. If the cancer had spread, then the whole organ would be scooped out then and there. He agreed to this plan, the biopsy was a thumbs-up, and he went home with half a thyroid. But a week later at his post-op checkup, the doctors saw signs the cancer could be spreading after all. Over the next weeks, Scharf received three contradictory opinions (remove, don't remove, and it's a toss-up), so he went with the safest course and had his thyroid removed for the second time.

The next day, Scharf was surprised by one more thing. The treatment plan called for him to take a radioactive iodine tablet just in case there was any cancer left. Then his doctor explained there were "some precautions" with this pill. For the next week, he couldn't be around anyone who was pregnant. He couldn't sleep next to his wife. He had to change the bedding every night and wash it multiple times. And he needed to flush the toilet three times because his waste would be radioactive. Basically, this pill would turn him into a human radiation hazard, and the doctor said the best thing he could do would be to hole up in a motel for a week by himself and then all would be well.

Scharf gaped at this litany, which reminded him of nothing so much as one of his early sixties grade-school lessons on surviving nuclear attack. He finally found his voice and said, First of all, Doc, I didn't know about this "one more thing" before the surgery. And second of all, I can't move out for a week, because I have an ailing wife and ailing dogs and ailing horses and I need to take care of them. Not to mention what might happen to the room cleaners at the motel. What if one's pregnant? What about the next guest who checks into Motel 6 and gets exposed? So, thanks but no thanks, I think I'll pass on becoming radioactive.

The denouement to Scharf's tale: He must be checked every year for thyroid cancer. He has been all clear since the surgery, nine years and counting. Recently, his doctor told Scharf he had made a good decision in declining to become a walking bar of uranium. But Scharf figured this meant he probably could have safely chosen to keep his half thyroid, too. Now he has to take those darn meds every morning for the rest of his life for no good reason, when he'd rather be frying up some bacon and eggs.

He usually finishes the story with a chuckle.

That pleasant little laugh is the most impressive part of this (and most every) Scharfian tangent: there is no bitterness or outrage. These stories are told with humor and a sort of bemused geniality, with a slight, wry half smile that makes you think of a favorite uncle. It's clear why he has achieved his reputation for coaxing crooks to confess. Where does this equanimity come from? Well, he says, look at it this way: things could have been so much worse. What if skipping the radioactive iodine pill had turned out to be a *bad* decision? He considers himself lucky. So this is how Scharf starts each day, with a dosing that is simultaneously ritual, remedy, and reminder.

After the thyroid meds, it's time to feed Flop, the tuxedoed black cat named for her habit of flopping down bonelessly in front of people to get their attention. She is an outdoor cat, given to him by his original cold case partner, and now and then Flop greets Scharf with the gift of a dead mouse. This day, though, the cat merely offers her dad the ritual flop. As he puts the bowl down on the back porch, Flop rises and rubs against one of his legs, leaving behind a generous deposit of cat fur on his pants, then takes a dainty bite of breakfast.

Next, Scharf walks to the small barn he built himself, stooping to pick up a hay bale for Chance and Penny, the Scharfs' rescue horses. He leans back to balance his center of gravity as he lugs the bundled brown stalks to the pasture. Cold case detective Scharf has lost the lean, rakish look of patrol deputy Scharf. What's left of his hair is now salt-and-pepper gray, and there's a prominent paunch at his beltline, which he attributes to his love of classic Pepsi, the caffeine of choice for many coffee haters. The farmwork keeps the former city boy strong, though. And the pale blue eyes are the same as ever, watchful and expressive, marksman's eyes.

He drops the bale with a *thump* and a groan, tears part of the bale away to make it easier for the horses to bite, then fetches some handfuls of bright green alfalfa. Chance and Penny have free rein of the Scharfs' 5.7 acres. Before they married, Laura lived on the property in a mobile home with her two young children from a previous marriage. Scharf moved in, they saved up for five years, and then they bought their three-bedroom house from a prefabricated home builder. Rheumatoid arthritis ended Laura's riding days and then her working days, and her mobility is limited by constant and debilitating pain. But she still enjoys watching from the window as the horses graze and romp like huge dogs.

Once Scharf has told Penny how pretty she looks, and he's finished cooing and encouraging Chance to take his daily pills, it's time to feed the pugs. He has to cajole Cheyenne to eat so he can administer her daily dose of insulin. Her appetite isn't great, so it's a daily battle for Scharf, and he worries about her constantly. The energetic Harley sleeps cuddled with Laura every night, but Cheyenne is Scharf's favorite. He has outfitted her with a silver tag bearing the name "Killer" with a heart.

Next, he returns to the house, makes a snack for Laura to eat when she gets up, and fills up her CPAP breathing machine with water to keep the airflow humidified. For a long time she took a powerful med for her arthritis but had to discontinue it when she got lymphatic cancer. She beat that but has had respiratory issues ever since. Finally, Scharf rubs CBD lotion onto her lower back, the only relief that seems to work for her constant pain.

His morning chores complete, Scharf dresses for work, ties his tie, and then heads to the office. A half hour later he pulls into the dim county garage redolent of decades of old car exhaust. He walks to his office along employees-only corridors lined with piles of discarded office machinery, old furniture, unwanted lamps, and battered filing cabinets. It feels more like a stroll through a dusty museum storage room or disused attic than a passage to the courthouse and the sheriff's department.

ON THE MORNING of August 20, 2010, Scharf already felt tired before reaching the elevator. He woke tired. He cannot fall asleep until after midnight or one a.m., no matter how hard he tries. It's the cold cases that do it, he supposes. It's the loose ends. It's the unanswered questions. It's the memory of Bill Van Cuylenborg's search for his daughter, and the frustration Scharf felt at his inability to close that case and so many others.

Last night, there was one more thing keeping him awake: excitement. He'd be going after another suspect in Tanya and Jay's case today. The cold case deck had done its work again, generating news coverage in Canada, with the king of hearts as the focus. A tip had come in.

And this one was different. This suspect had a proven connection to the case, had even admitted being involved, and yet he had eluded police for decades. Investigators on the case had been waiting for this break since Christmas 1987.

They had called him the Taunter. But now Scharf knew his name, the sender of those Christmas cards, the man who claimed credit for the murders of Tanya and Jay.

14

The DNA Blues

August 20, 2010
Seattle, Washington

The tip came in a Canadian news story on cold cases that showed images of the Taunter's lurid letters to the Cook and Van Cuylenborg families, including the scrawled "Hallelujah BLOODY Jesus" declaration.

A day after the article appeared, someone contacted the police in Canada by an anonymous email to say they recognized the handwriting. The email included a first and last name for the Taunter, as well as a possible old address.

Scharf's first step was to root through the old case reports to see if that name had ever popped up before. There were thousands of pages, three bookshelves of loose-leaf binders, and he had a group of volunteers who had helped him sort through it all—a retired judge and a retired probation officer among them. They found the same name in an old report. It had been there since 1988 but had not gotten any follow-up.

This disgusted Scharf, but it did not surprise him. He had found viable suspect names buried in old reports of other cold cases that had never been checked out, though dropping the ball on the author of these letters was a pretty big fumble. In high-profile cases, with hundreds of tips flowing in and the pressure for results at maximum, the investigation would have operated in triage mode, tackling the most promising leads first and setting the others aside for later. Sometimes later never came.

The Taunter had mentioned Ann Rule in one of his cards to Gordon Cook, and a detective at the time had contacted the crime writer and former Seattle cop on the off chance that she knew something.

Rule had told him yes, she knew who this was, and supplied the Taunter's real name. But whatever follow-up there had been, if any, it had stopped short of finding the man who the most popular crime writer of that era had named for them within months of the murders. So Scharf mounted a thorough effort to trace the man, with the help of a Skagit County detective, Tobin Meyer, the first investigator from the neighboring county to express a desire to partner on the case in many years.

They learned the man, a former teacher, had left his profession, lost his home, and had been moving from city to city in Canada, Washington, and California, staying at shelters and homeless encampments. The two detectives traveled together to the Taunter's last known address in Vancouver, British Columbia. They were eight months too late. But their target left a trail of forwarding addresses and other clues, and they eventually traced him back to Washington and then to their own backyard, downtown Seattle. This heightened Scharf's suspicions, given that the Canadian travelers were headed to the same place when they disappeared. Downtown served as a major destination for the homeless of Washington—shelters and services were concentrated there. The world-class public library main branch downtown was a big draw for the unhomed, with its large collection of newspapers from around the world and its banks of free internet-connected computers for public use. The Taunter's Seattle stomping grounds were a ten-minute walk from the ferry terminal where Jay and Tanya would have arrived.

The library was Scharf and Meyer's first stop. They had found a fairly recent photo of the man, and the library security officer said in an offhand way, "Oh, sure, he's in here all the time."

There was nothing offhand about Scharf's and Tobin's reactions. They were jubilant, excited. A major suspect in the case was within reach.

Scharf knew his predecessors had doubted the confession contained in the Taunter's letters. The detective knew it could be a hoax and a dead end, a twisted brag. But then again, Scharf could be on the verge of clearing an epic cold case. His very first cold case. He might finally have answers for those families.

The next morning, they had him. As they hung around the newspaper collection, Scharf spotted the Taunter. He and Meyer exchanged a look, then walked up to the man, whose eyes at first stared back blankly and blandly at them, then moved to the badge on Jim Scharf's belt.

Then the Taunter burst into tears.

Yes, he said, he had written the letters. He had sent those cards. That had been a terrible thing to do, he said, and he had been sorry for doing it for many years. He knew the police would catch up with him someday.

The Taunter seemed more relieved than fearful. Scharf ended up feeling sorry for him. He had endured an abusive childhood, they learned, one he thought he had overcome. He told the detectives he had taken a job teaching in a private school in Canada, but when it became known he was gay, some parents objected and he was let go. This was many years ago, he said, when discrimination was rampant and protections were scant, and he spiraled into despair and hatred of his own countrymen. The letters were his way of lashing out during the lowest point in his life.

"But I had nothing to do with killing those poor kids. Nothing."

Scharf and Meyer believed him. They would not be solving Tanya's and Jay's murders this day. The Taunter was in his seventies by then, a broken man, lonely and still struggling with his mental health. The statute of limitations on the harassing letters had long since expired, so there would be no charges for that. He voluntarily provided an oral DNA swab and, as Scharf felt certain they would, the results cleared him of any involvement in the crime.

Identifying the Taunter resolved a lingering mystery. Scharf had something to tell Tanya's brother and Jay's dad. It was a triumph of sorts, he told himself. It just didn't solve the case.

The same thing had just happened with the latest two tips in the case: one tip about a man who bragged about killing a couple in Sno-homish County, and another tipster who believed her ex-husband in Las Vegas had killed Jay and Tanya. The ex-wife sent Scharf an enve-lope her husband had licked. Scharf, meanwhile, tracked down the braggart to a street corner in Seattle's Queen Anne neighborhood, where he tried to sell copies of an alternative newspaper, *The Stranger*, for a dollar apiece to shoppers leaving a Trader Joe's store. Scharf posed as a shopper with his arms full of groceries and asked the suspect to put a copy in an envelope for him so he could send it to his brother. The man, lured by the five-dollar bill Scharf waved in front of him, did so happily and even licked the envelope for him, another DNA sample for the lab. Both were ruled out as Tanya and Jay's killer.

So it went, Scharf thought, in this golden age of DNA matching. The ability to identify or exonerate a suspect with trace genetic evi-dence was truly a miraculous advance, more accurate and powerful than every other forensic technique used to identify a criminal. Scharf should know: he had solved numerous cold cases with DNA that could not have been cleared any other way. But not Jay and Tanya's. For them, the DNA for every suspect always came back with the same finding: Scharf had the wrong guy.

The cold case detective joked that he had the DNA blues.

THERE HAVE BEEN three distinct phases in this golden age of using DNA to solve criminal cases, with each phase more capable and revealing than the one before. When Scharf was stalking the Taunter at the Seattle library or standing outside a Trader Joe's to get a suspect's DNA, the second era of forensic DNA was well under way. It was not enough to crack Tanya and Jay's case. For that, Scharf would have to wait for—and help launch—the dawn of the third DNA age. But the second era was still a vast improvement on what had come before.

The first age, the age of DNA fingerprinting, dates back to September 1984. That's when thirty-four-year-old British scientist Alec Jeffreys had what he later described as his "eureka moment" while looking at X-ray photos of a DNA experiment gone badly wrong. He was working to chemically extract small sections of the DNA molecule, then visualize the pattern of pairings of structures in our genes called "alleles" on X-ray film. He had hoped to uncover genetic similarities between family members with the same hereditary diseases. Instead, he found dissimilarities no one expected with closely related people.

Rather than tossing aside an unsuccessful experiment, he realized he had stumbled on a breakthrough—a new way of identifying one human from the next. These tiny areas of dissimilarity on the massive DNA molecule provided a simple genetic snapshot unique to every human individual. He had discovered the equivalent of a fingerprint in our genes.

But this was better. Real fingerprints have to be judged by human interpretation of smeary partial prints on murder weapons and steering wheels and beer bottles. It is often as much art as science. Errors and bias can creep in. Matches can be missed. DNA fingerprints, by contrast, are cut and dried, no human judgment required.

While mapping the entire human genome was a daunting project that would take billions of dollars and several decades to complete, Jeffreys had jumped the line with a cheap shortcut. The small sections of the genome he examined, called STRs (for short tandem repeats), are basically fillers and separators within the DNA double helix. They don't actually do anything other than act like molecular bubble wrap, dividing more complex sections of DNA that code human traits—eye color, nose shape, gender, and the rest of our genetic blueprint. In other words, STRs, like physical fingerprints, tell you nothing about the nature and qualities of an individual, yet they are perfect for distinguishing that same person from every other one in the world.

Jeffreys's theory was soon tested in the real world in 1987 to solve the murder and rape of two fifteen-year-old girls in Leicestershire County in the English Midlands. This method, for which Jeffreys was knighted, first freed the jailed prime suspect, a learning-disabled seventeen-year-old badgered into a false confession. Then DNA matching led to the arrest of the true culprit, who promptly confessed and pleaded guilty. A revolution in criminal forensics had begun . . . almost. Because there was no trial, the reliability of DNA evidence remained untested in a court.

That changed later the same year with the first criminal trial in the world to use DNA fingerprinting. A jury in Florida convicted Tommie Lee Andrews, accused of multiple rapes and robberies—and the main evidence against him was his own DNA. The verdict and its endorsement of DNA fingerprinting as a reliable tool of the justice system came in November 1987, just a few days before Jay Cook and Tanya Van Cuylenborg died.

But there was a fundamental limitation to this first age of DNA matching: the police first had to identify a suspect through other means in order to have something to compare with the crime-scene DNA. This early iteration of DNA fingerprinting was a confirmation tool, not a search tool.

In the Florida case, the rapist had been caught fleeing the scene of his final crime. In England, Jeffreys's first case got around this limitation in a more extraordinary way: the small-town murders of two young girls so aroused public outrage and fear that the area's entire male population, 5,511 people, voluntarily gave DNA samples to the police. Jeffreys had to compare each of them to the crime-scene DNA, one by one, to find the culprit. This was a once-in-a-lifetime event, though. Authoritarian regimes might order an entire population of law-abiding citizens to surrender their DNA, but that would never fly with the American public or pass constitutional muster. So in the United States, the FBI chose a different solution for searching the DNA of an entire population: they set out to build a DNA database.

Because DNA fingerprints can be reduced to alphanumeric expressions—computer code—they can be compiled in a searchable digital index. With that, detectives could compare crime-scene DNA to thousands of suspects far more quickly and accurately than a one-by-one comparison by human eye. The Constitution didn't allow a whole American village to be DNA fingerprinted, but its lawbreakers were fair game. If their fingers could be printed, so could their genes.

With that, the second age of DNA matching began, the age of the database, with the birth of a system eventually called CODIS. The Combined DNA Index System grew slowly from a few thousand DNA profiles of arrestees and convicts to several hundred thousand, then to millions by the end of the century. It has been used to identify more than a half-million suspects since its inception.

In 1994, the DNA of Tanya and Jay's killer was uploaded for the first time to CODIS's forerunner, and again seven years later to the system still in operation today. Since then, that DNA has been continuously compared to all the criminals in CODIS, as well as unknown DNA from other crimes nationwide.

No match has ever been found.

And so the limitation of the second age of DNA matching continued to plague Jim Scharf. The Taunter wasn't in the database. Neither was the Trader Joe's newspaper guy, nor the man in the Walmart. Nor was Tanya and Jay's killer. This didn't mean he hadn't done other violence, Scharf knew. He just hadn't been caught. To CODIS, though, that meant he was invisible.

Forensic scientists accepted this blind spot as a given. CODIS was an amazing tool, nonetheless, but DNA comparisons could only go so far. Or so they thought.

But they were wrong.

It would take something entirely different to address that critical shortcoming and to launch the third age of DNA matching, and it came from a most unlikely source—one that had nothing to do with law enforcement or crime detection at all.

It came, instead, from our personal obsession with our roots.

AROUND THE SAME time that Alec Jeffreys's DNA fingerprinting debuted in a Florida courtroom, and Jay Cook and Tanya Van Cuylenborg planned their first and last trip abroad together, another kind of DNA breakthrough created an international sensation. Her name was Mitochondrial Eve.

New Zealand–born biochemist Allan Wilson had become obsessed with finding the common ancestor of every living human—the great, great, many thousand times great-grandmother of us all. He was not searching for old bones in caves or fossils in remote outposts. He was searching for ancient DNA—inside us all. The clues to our origins live in our contemporary genes, the traces we inherited from our earliest ancestor.

In his search, the MacArthur genius-grant winner focused on the chain of heredity embedded in our mitochondria, unique structures in all plant, animal, and human cells that convert nutrients into energy. Mitochondria are unique in another way, too: they carry their own DNA, whereas most of our genetic code resides in another structure, the cell nucleus. Unlike nuclear DNA, which is passed on to children in roughly equal measure from both mother and father, mitochondrial DNA (mtDNA) is passed on only from mother to child.

It also mutates at a very slow and predictable rate. Your mtDNA and your ancestors' from ten generations ago are very likely identical. But over very long periods there is variation, dating back to prehistoric times, when there was no global travel or mating between regions. That means there are measurable differences in our mtDNA linked to geography. This is how consumer DNA tests today can tell that someone's great-great-great-great-grandparents hailed from Ireland or Palestine or Kenya. Wilson was the first to use the principle, turning it into a kind of molecular time machine that could unwind all those variations back across generations to their convergence point both genetically and geographically—our female common ancestor. Which would be every living human's common ancestor.

Wilson found that the female from which every modern human was descended lived in Africa 150,000 to 200,000 years ago, which the world learned in October 1987, when the magazine *Science* published an article entitled "The Unmasking of Mitochondrial Eve." The very concept that such a thing could be knowable, the scientific personification of the notion that we are, in fact, all brothers and sisters, captured the public imagination as no other scientific advance had since the first moonwalk.

This was the birth of genetic genealogy.

MtEvc is a theoretical individual—there is no burial site, no name, no bones—but her genetic trace lives on inside us all. The discovery drove others to tackle research that had been stymied before this new genetic genealogy arrived on the scene.

Some of these new projects created sensations because they focused on more recent, known historical figures. One of the first outed a false pretender to the throne of King Louis XVI and Marie Antoinette. A similar study using male Y-DNA had a different outcome. Familial DNA comparisons proved that Thomas Jefferson had fathered the children of a woman he had enslaved, Sally Hemings. This vindicated the Hemings family's longstanding oral history, sweeping aside the denials and disputes from some of the third president's other descendants. This was a new application of Y-DNA matching, long a favorite tool of genealogists researching surname ancestry, tracing back last names into antiquity. Both surnames and Y-DNA pass from father to son, so this blending of old and new methods was a match made in genealogical heaven.

Public interest in old-fashioned genealogy, based on birth and death records, other archival documents, and oral family histories, had already been on the rise ever since the 1977 television miniseries *Roots*. Based on the Alex Haley book of the same name, the series finale was watched by 71 percent of households in the country. The trend was further fueled by the rise of the personal computer and the internet, which offered easy access to records and the ability to share documents and family trees instantly anywhere and with anyone. *Time* magazine declared 1999 the year of "Roots Mania," as 160 million messages a month were sent between the users at RootsWeb, an early, free online genealogy exchange and research site later acquired by the current industry leader, Ancestry. The article concluded that a hobby once the realm of "self-satisfied blue bloods tracing their families back to the *Mayflower*" had become the new passion and connector of a multicultural society.

The first home consumer DNA tests came to market a year later. The tests initially focused on Y-DNA and mtDNA, but they grew in power and reach over time. They were designed to help find people with similar genes—relatives, both close and distant. It was the consumer version of Allan Wilson's search for Eve, except you could find relatives who were still alive. And you didn't just find one match. You found dozens, sometimes hundreds. Genetic genealogy is designed to be a tool of inclusion—the opposite of DNA fingerprinting, which found the one person on the planet who matched a crime and dismissed everyone else. Criminal forensic experts preferred the exclusion of DNA fingerprinting and saw no value in its reverse.

But a few people began to wonder what might happen if you could combine both technologies. CeCe Moore, who found the identity of Baby Alpha Beta with genetic genealogy, was among the first to ask that question. Might that combination of scientific opposites lead to a new way of finding people?

15

The Tool of Inclusion

Before genetic genealogy became CeCe Moore's obsession, life's work, and pathway to fame, it was something else entirely to her: an engagement gift.

Back in 2002, Moore was acting full-time, juggling work, marriage, and parenthood. Grabbing something at the mall for her niece's engagement would have been the simplest choice, but Moore wanted to give something personal and creative. She remembered being fascinated as a child with a self-published book on the family's ancestry that her aunt had sent her and thought her niece might like something similar. But she wanted it better, updated. What if Moore could create a family-tree album that went beyond the old photos, records, and oral histories for a deeper ancestry, the kind of information people were obtaining with those new consumer DNA tests on the market?

The off-the-shelf mitochondrial and Y-DNA tests available at the time really did provide new information about the origins of the Moore family line, revealing a large number of possible relatives with

European roots. Moore's fascination with the hunt for these long-lost potential family members kept her on the computer for many hours. She was still at it long after the engagement gift was delivered and her niece's wedding had come and gone.

Moore's hectic workdays were divided between acting gigs onstage in regional light opera and musical theater, as well as regular bookings for television commercials. Then there were her weekend appearances for the Mattel toy company at store openings and toy fairs where, with her wavy, long blond hair, she impersonated the toymaker's signature doll, Barbie. Her downtime, however, increasingly went to genealogy.

Textbook laboratory science might not have been her forte during a college career focused on music and drama, but this genetic genealogy was different. This was assembling a puzzle, and Moore *loved* puzzles. Back when she still had free time, it wouldn't be unusual for her to have multiple complex jigsaw puzzles spread out around the house, all at various stages of completion, all going at once. She recognized that putting together family trees with both genealogy and genetics provided her with the ultimate puzzler's challenge. Seeing the branching complexities of a family tree take shape brought her the same familiar, deep satisfaction she felt after snapping the final squiggly piece of a tabletop jigsaw into place.

And something else familiar to inveterate puzzlers was also the same: the maddening frustration Moore felt after realizing that some of the final pieces were missing from the box after many hours of work. Genealogists call these "brick walls"—gaps in the family tree for which there are no record, no oral history, no old photos, and few definitive clues in the mitochondrial and Y-DNA databases available at the time.

But that changed dramatically in 2007 when 23andMe launched the first consumer-level DNA tests that looked at all twenty-three chromosomes that humans inherit in equal measures from mother and father—referred to as autosomal DNA. Unlike those bubble-wrap STRs used by DNA fingerprinting, autosomal tests inventoried SNPs—single nucleotide polymorphisms—commonly referred to as "snips." These are the coding parts of the DNA molecule, where your unique traits are determined. The company's initial emphasis was on using genes to identify a person's ethnic and geographic origins and predisposition for certain traits and diseases, ranging from such relatively benign conditions as baldness and back hair to life-threatening conditions such as high blood pressure, Alzheimer's, and heart disease.

These health products were criticized by some scientists at the time as "genetic astrology," and the federal government eventually intervened. The U.S. Food and Drug Administration ruled that the test kits qualified as medical devices that had to be clinically tested and approved by the agency, and sales of the health portion of the testing process were suspended in the United States in late 2013. The FDA approved the kits, and sales resumed in 2015.

Meanwhile, out of concern that private consumer DNA test results could be accessed by unscrupulous individuals and corporations in order to discriminate against people seeking insurance or jobs, Congress had passed GINA—the Genetic Information Nondiscrimination Act, with the intention of protecting U.S. citizens from genetic prejudice.

Meanwhile, 23andMe found another way to grow its DNA testing business by rolling out a new product: a supercharged DNA genealogy tool called Relative Finder. Each customer got a long list of distant cousins and other even more distant possible relations. Moore used these lists to find records, newspaper obituaries, birth notices, and all the other archival sources that genealogists dig through to construct a web of connections and insanely detailed family trees. The DNA lists provided her with avenues to search that were simply impossible for old-school genealogy methods alone.

Demand grew so rapidly that a competing company, FamilyTree-DNA, came out with its own autosomal home test and relative finder.

This wild popularity had an unanticipated effect: as the number of people who took consumer autosomal DNA tests grew from thousands to millions, the power and reach of genetic genealogy searches also grew, mirroring the growth of the FBI's database. But the genealogy databases could do something the criminal system was specifically designed to avoid: one relative provided all an adept searcher needed to identify hundreds of other people whose SNPs held inherited similarities. This meant, in shockingly short order, a majority of Americans could be found through these consumer databases whether they had taken a DNA test or not. Though few understood it at the time, the consumer databases' potential had far outstripped the FBI's system with its method of finding a single exact match.

More than anything else, it was these relative-finder tools that sold Moore on the power of genetic genealogy versus the old-school scrapbook methods of hobbyists for generations. She cajoled forty of her relatives to test their DNA in order to illuminate the farthest and most obscure reaches of her family tree. She joined the International Society of Genetic Genealogy and started swapping tips and techniques with a handful of other advanced early adopters in this emerging field. And Moore became a beta tester for both companies as they launched new and more far-reaching relative-finder tools.

Choosing the handle "Your Genetic Genealogist," she next began blogging in the spring of 2010 about her own family genealogy projects. One of her more dramatic posts recounted how she used her family-tree-building skills to help her brother-in-law, John, learn he was a descendant of Thomas Jefferson and Sally Hemings. The ex-president and ex-slave were John's great-great-great-great-grandparents on his mother's side. Moore's moving account recalled how deeply this revelation affected her brother-in-law, who had known little of his family history and who made a pilgrimage to Monticello afterward with his relatives to explore these roots.

But a cautionary note colored Moore's post, as she had found a disturbing flip side to the story: family lore passed on to her brother-in-law growing up was that they had some Native American ancestry, not African American. Moore was learning—and telling the world in her blog—that genetic genealogy was not only about discovering the names and stories of long-dead ancestors. It was also a means of exposing the mistakes, fables, cover-ups, and false narratives that can be embedded in our stories and beliefs about our roots—within families as well as communities and nations. Genetic genealogy was revealing important truths about the living, not just the dead. It could expose secrets and lies that had been obscured by time and bigotry as easily as flipping a rock.

Her brother-in-law's story resonated with Moore's audience. Some of them had uncovered family secrets of their own—wonderful for some, devastating for others. Messages began coming in through her blog from newbies and experienced genealogists alike. They had spent years assembling what they believed to be accurate family trees based on government records, grave registries, news archives, old letters and photos, and relatives' recollections and anecdotes faithfully passed down across generations. Now they stared in bewilderment at DNA test results that overturned what they long believed to be bedrock truths about family and identity. They were finding out their fathers were not their fathers. That people they called Aunt or Uncle all their lives were not actually biologically related. That family lore was, in fact, family fiction. Some learned through a mail-order test that they had been adopted, and they were now desperate to find their birth parents. All were reaching out to Moore for help in using their DNA results to find the truth.

Moore began to puzzle out how to adapt her existing genealogy techniques to tackle these very different questions. In all her previous projects, she started out with information most people take for granted: she knew the names and relationships of close family and their parentage—aunts, uncles, first and second cousins, grandparents, and, of course, parents and siblings. She used this basic information and DNA tests from these family members to find links to ancestors from earlier generations and more relatives. In this typical genealogy search, Moore reaches into the past. But now she wanted to help people who didn't know their closest living relatives, who couldn't name their biological aunts or cousins or parents. She would have to use genetic genealogy to flip the process and seek out the closely related living. It was a different way of thinking about the purpose of genealogical research, a kind of family-tree building in reverse. But she saw no reason why it couldn't work.

In 2012, that belief was put to the test by a distraught family whose DNA tests showed their adult daughter was biologically related only to her mother, not to her father. Ashley and her parents had been living a lie for more than twenty years. Now they wanted answers.

Moore had been shocked at first by how often DNA tests pierce this particular secret, one so common that genetic genealogists have developed a polite, discreet acronym for it: NPE. Depending on who's talking, this can stand for "non-parental event," "non-paternal event," or the most clearly descriptive, "not the parent expected." There are no definitive studies on how common NPEs are, but the best estimates put it at 1 or 2 percent of every generation, meaning that between one and two out of every hundred people in your age group believe the wrong people are their biological parents.

NPEs most often are attributed to marital infidelity, although there are other explanations as well: secret adoptions, sexual assault, incest, sperm and egg donation, and hospital switched-at-birth errors have all caused NPEs. But the family who contacted Moore had the most unusual tale yet: Ashley had been conceived with the help of fertility treatments that were supposed to have used her father's sperm.

Somehow, the fertility clinic in Utah had made a switch—a swapped-before-birth error. They never would have known, or even suspected there was a problem, had they not decided to do a consumer DNA test just for fun. Devasted by the test results, the family wanted Moore to help figure out what happened.

Moore became obsessed with finding answers for Ashley and her family. She, her laptop, cell phone, and coffee cup would convene every day on the big beige couch in her living room, where she lost herself in work for eight, ten, or more hours nonstop, combing through family trees and old newspaper archives. The irony of her work is that, the more she becomes enmeshed in other people's histories and tragedies, the less time there is for her own family, for exercise, for getting off the couch—for her own life.

Moore found on Ashley's DNA test match list a biological cousin related to the daughter but not related to her mother. This had to be a relative of the source of the sperm—the daughter's unknown biological father. After a long and convoluted archival search, Moore traced the cousin's family tree back and forth through time and across multiple family branches until she finally found the person who supplied half of Ashley's DNA, a man named Thomas R. Lippert. And that was when she learned this had been no accidental switch at the fertility clinic. It had been done on purpose.

Lippert had worked at the Salt Lake City fertility clinic Ashley's parents had used. A convicted kidnapper who had served two years in prison in the 1970s, he inexplicably had been hired to staff the clinic's front desk and also served as a lab technician. He was the tech who had retrieved the sperm sample from Ashley's dad that was supposed to be used that day. Instead, he substituted his own.

Lippert died in 1999, long before the scandal surfaced and before the forty-nine-year-old lab tech could be questioned. The clinic had closed down, too, its records destroyed. No one knows if Lippert made other switches, and the subsequent investigation was hampered by the fact that he was also an authorized sperm donor used by clinics around the country. His glowing biography, shown to parents desperate to have an exceptional child with DNA from a supposedly exceptional donor, somehow neglected to mention his record as a felon who did hard time or the fact that he threatened to kill his wife if she ever tried to leave him.

The university eventually apologized to Ashley and her parents, but the extent of the damage done remains unclear. Lippert may have "fathered" hundreds of children around the country under false pretenses. Moore believes this showed a dire need for greater regulation and patient protection in the fertility industry. Similar scandals involving fertility fraud have been unearthed by the advent of DNA testing, often involving the fertility doctor himself substituting his sperm for the father's or the selected donor's.

Meanwhile, Moore gave up acting to become a full-time genetic genealogy consultant. The field had no course of university study, no degree or certification, no official standing. But she did join a small group of pioneers on the same path. They saw themselves as leading a citizen-science movement—grassroots, crowd-sourced, volunteer-driven scientific inquiry, sharing techniques, creating ethical guidelines, and making rapid advances few outside their small group saw or understood. The consumer DNA companies had created a tool that was largely recreational and immensely popular, but this cadre of new genetic genealogists wanted to use it for an entirely different purpose. The rest of the world might mistake what they were doing for the quaint hobby of family-tree research, the biological equivalent of antique hunting. But Moore felt she was onto something that could do so much more.

For a time, she teamed up with a group of "search angels"—genealogists who specialized in records-centric methods of helping adoptees and their birth families reunite. Eventually Moore and the search angels cofounded the nonprofit DNA Adoption. She shared with them her techniques while they educated her on how to research the closed and privacy-shrouded world of adoption records. Later she moved on to her own DNA Detectives Facebook group for helping adoptees, foundlings, and people with unknown parentage and other mysteries on their family trees.

Moore used the same reverse-family-tree-building techniques to help an amnesiac named Benjamin Kyle find his true identity and roots, and to reunite Baby Alpha Beta with her birth family. In 2013, Moore became a staff genealogist for Harvard University professor Henry Louis Gates Jr.'s PBS show, *Finding Your Roots*. There she displayed her genetic genealogy methods to a national audience, which would eventually include Scharf. Through Moore's work on the show, rap artist and actor LL Cool J learned that his mother had been adopted. Moore then introduced Cool to the ninety-year-old biological grandmother he never knew he had.

This level of exposure and media coverage drove rapid growth in her DNA Detectives group, and volunteer genealogists around the country signed on to help the burgeoning membership find lost relatives. By then it had become obvious to Moore that other sorts of mysteries could be solved with genetic genealogy, not just NPEs. Her method of identifying an amnesiac and a foundling by testing the DNA in their saliva and then uploading their profiles to a genealogical database would be exactly the same if she tried to identify an anonymous killer or rapist who left behind blood, semen, saliva, or even skin cells under the fingernails of victims who fought their attackers. Cases like Baby Alpha Beta were her proof of concept: Moore had perfected a method to solve previously unsolvable crimes and cold cases. And she knew it.

Genetic genealogy, she had come to believe, was the source that never lied, never faded with time, never forgot. It was the forever witness.

But her attempts to convince others in the industry about the possibility of expanding into crime fighting did not go well. Her genetic genealogist colleagues worried that cooperating with the police would spook people about getting tested, and the power of their new DNA databases would stagnate amid fears about privacy and government intrusion. The insular genetic genealogy community was sharply divided on the question, with some open to working with police on solving violent crimes and others arguing strongly that it would be a misuse of private DNA test results without permission from the people the data belonged to.

Her contacts at the big consumer DNA companies were even less enthusiastic about allowing DNA from unsolved violent crimes to be uploaded by law enforcement. They didn't even want to do a poll to see if their users would be open to it. They said their business model depended on maintaining a position of trust with their customers, with an absolute guarantee of genetic privacy. Customers wouldn't feel their precious genetic information was very private if cops were rooting around looking at it, Moore was told.

She was disappointed but not discouraged. By then, Moore had become an influencer with their customers, so she felt she could try again down the line.

At conferences on genetics and DNA in law enforcement, she tried to convince police detectives, forensics specialists, and researchers in the field of human identification that far more could be done with genetic genealogy than learning who your great-great-grandma was. If they could work together to launch a third age of DNA crime fighting, Moore believed, there would be no place left for criminals to hide.

But the forensic experts would just shake their heads politely, just as dubious as the consumer DNA companies. The forensics community already owned a tool that had revolutionized crime fighting, DNA fingerprinting, and a massive database that had been years in the making, with every police agency and crime lab in the country participating. They were not about to switch. What could leading DNA scientists who had been doing cutting-edge crime lab work for years, even decades, learn from hobbyists building family trees with sixty-nine-dollar consumer DNA kits bought on Amazon?

16

Precious Jane Doe

Had Jim Scharf met CeCe Moore at one of those forensics conferences, their work together on cold cases would have begun years sooner, the detective later lamented. He would have seen the value of her work in a heartbeat. But the California-based genealogist and the Washington-based detective had not yet met or heard of each other. Years would pass before luck would bring them together on Jay and Tanya's case.

Even so, Scharf was always on the lookout for new approaches to solving old cases, and he had read about early forms of genetic genealogy. He had tried Y-DNA in a few cases, because there have long been databases of male gene profiles that are public and can be searched by anyone for matches. But he had found their utility limited. All a detective got back was a list of last names that might match the killer. Running down such lists to see if any of the names were the right ones had never proved useful for Scharf. He knew Y-DNA had occasionally produced a useful tip for other detectives that helped narrow an existing list of suspects. But there have also been cases where Y-DNA led to false accusations against the innocent. None of this gave him a clue about the sort of power behind the genetic genealogy Moore and a handful of others had begun to practice.

Then Scharf read about a case in which a group of genealogy volunteers at the nonprofit DNA Adoption used a new kind of DNA analysis. With it they had figured out the true identity of "Lisa," a woman in her thirties who had no clue about her birthplace, the identity of her family, or her real name.

Lisa had been abandoned at a Southern California campground in 1986 when she was five years old by a man she and others assumed had been her abusive—and elusive—father. But he had been her kidnapper, not her dad, and his name was an alias. The mystery endured until the leader of the volunteer effort, a retired patent attorney named Barbara Rae-Venter, had Lisa take several consumer DNA tests. Then she used the relative-finder tools to locate one distant cousin after another. The team eventually found more than two hundred relatives. Most agreed to test their own DNA, deepening the data and widening the genetic net. At last what had become known as the Lisa Project found her birth family in 2016. It had taken twenty thousand volunteer hours to reverse engineer Lisa's family tree, but she was finally reunited with a grandfather and other close relatives and learned that her given name was Dawn.

Meanwhile, clues uncovered in the process helped police determine that kidnapping a child had been the least of her fake father's crimes. That man, who had gone by so many names the police called him the "Chameleon Killer," was a serial murderer. His victims likely included Lisa's biological mother, who had disappeared around the time of Lisa's abandonment. The Chameleon Killer had died in prison in 2010.

The Lisa Project story electrified Scharf. He realized that this sort of genetic detective work was different than just getting a Y-DNA list of last names. This brand of genetic genealogy got tangible, actionable results. Yes, the twenty thousand hours were daunting, but other cases couldn't all be so complex or labor-intensive. He figured anyone who could use DNA to identify a nameless person, someone who had been a living Jane Doe for thirty years, just might be able to identify Tanya and Jay's nameless killer after thirty years as well.

He called Rae-Venter the next day. But he was in for a disappointment: she told him that the consumer genealogical databases did not allow crime-scene DNA to be uploaded into their databases, so she wouldn't do suspect searches.

Before he could object, she said that didn't mean they couldn't work together. She could do genealogy searches for murder *victims*—if he had any that needed identifying. He did, of course. Every cold case detective does. Scharf wasn't going to give up on finding a way to use genetic genealogy on Tanya and Jay's killer and his other cold case murder suspects. But for now, he eagerly accepted Rae-Venter's offer and told her about Precious Jane Doe, his most unusual cold case, a case where the killer had been caught years ago but the victim remained a mystery.

THEY FOUND HER in the woods ten years before Jay's and Tanya's murders. Blackberry pickers stumbled on a young woman's body in the late summer of 1977, not far outside of Everett. She had been shot seven times in the head, and she had been in the thorny bushes far too long, her face unrecognizable from decomposition, though her shoulder-length sandy hair, neatly parted in the middle, appeared remarkably and jarringly pristine.

She wore a summery striped tank top, cutoff jeans, a yellow Timex on her wrist, and sneakers with no socks. She died with a half pack of Marlboros and seventeen cents in her pocket. Police found no identification.

An anthropologist who consulted on the case put her age at fifteen to twenty, most likely between sixteen and nineteen. She had been tall, five ten or more. Investigators figured her for a runaway, but they found no police reports or calls matching her description—not locally, at least. She became "Snohomish County Sheriff's Case No. 77-17073—Jane Doe" the day she was found and stayed that way for forty-three years.

It was a rare case where the police identified the culprit but not the victim. The killer had called a friend in a panic and said, "I just shot a hitchhiker and I need you to come help me move the body." The friend said he didn't really want to do that. Instead, he went to the police. And just like that, they had the name of the killer: David Marvin Roth, a gangly twenty-year-old who was so angry at himself for what he had done, he had taken the same .22 rifle he had used to murder the hitchhiker and shot his car full of holes.

Roth had actually been arrested for that, drunk or high or both when he was picked up for killing his car, a firearms violation. But he was released before the homicide detectives tracked him down. It was the 1970s, after all, and records still traveled with the speed of paper, not computer code. By the time anyone realized they briefly had a murderer in handcuffs, Roth was in hiding on the Olympic Peninsula, that wild wood of Washington where so much impenetrable beauty and inexplicable mayhem collides. It took a year for the police to track him down, after which he promptly confessed and was sentenced to life in prison.

He admitted everything. He had spotted a pretty young hitchhiker while driving to Silver Lake to go swimming and pulled over. From the passenger seat, she spotted the old, lurid seventies-era biker bumper sticker he had slapped on his dashboard—"Gas, Grass, or Ass (No One Rides for Free)"—but she didn't say anything about it. When he invited her to go get a beer with him, she said great. He stopped at a market, bought some beer, then drove to an old World War II bunker site on the edge of town, a common make-out spot at the time, now long gone. After a few beers, when he asked if he could touch her, she said yes. He fondled her breast.

But she said no when he asked if she'd have sex with him.

He was seething at that but apparently hid it well. He excused himself, went to the trunk, and came back with a peacock feather. He bent down to peer into her car window and offered her the colorful feather and said, "Would you like this?"

She smiled and said yes, but when she reached for it, she saw what was in his other hand: a long bungee cord, which he wrapped around her neck in a stranglehold. He dragged her from the car and choked her until she stopped struggling. He was about to leave her when she started to thrash again, so he walked back to the truck, got his rifle, and shot her seven times in the head.

But, he swore, she never told him her name. And he never asked.

Twenty-six years later, he was paroled. And Jim Scharf, now in charge of cold cases, including the search for Jane Doe's name, was waiting to talk to him. Roth seemed anxious to make amends. He had taken anger management classes in prison, and he genuinely tried to help Scharf. But he had no more information to add that could help the detective identify the girl.

Scharf tried everything. Artist sketches. Facial reconstructions. He had put out press releases and worked with a volunteer to produce a video about the case. The volunteer, who worked with a nonprofit that helped families search for missing children, suggested Scharf come up with a better name than just "Jane Doe." That's when he decided she'd be "Precious Jane Doe" until he could find her true name, and he told the volunteer why.

"Here she is, this seventeen-year-old girl, she's got some older guy hitting on her, wanting sex. And she's able to say no. She was able to turn this guy's advances down. And then it ends up getting her killed.

Now I just want to give her name back. And give her back to her family. She's got to be precious to somebody, I think. And if no one else, she's precious to me."

Scharf had Precious Jane Doe's body exhumed thirty years after the murder to get a DNA sample. The grave had been waterlogged and decomposition was extreme, but a university lab was able to develop a partial DNA fingerprint. Scharf had it uploaded to the FBI's national database, which tracks the DNA of missing persons and their close relatives, as well as offenders. But there were no hits. The DNA was too degraded for developing the more complex SNP gene map Rae-Venter needed for genetic genealogy.

Scharf had one more shot: in the evidence room, he still had Precious Jane Doe's hair, which had never been buried. It was in surprisingly good condition. Scharf found a scientist who specialized in prehistoric human DNA who could extract a SNP profile from hair. He had done it for a forty-thousand-year-old Neanderthal body, a monumental achievement. But Precious Jane Doe was almost as tough a problem. It took Dr. Ed Green at the University of California, Santa Cruz, months to finally get a usable profile. At the time, this was only the third successful extraction ever of a SNP profile from hair.

Scharf turned it over to Rae-Venter and eagerly awaited results, but she was much in demand, and months passed. Scharf considered finding someone else to work on it, perhaps a volunteer who was learning genetic genealogy and was already consulting a bit on his cases, but then something else distracted him. He learned of another new way to use DNA to catch criminals, and it was almost too outlandish to believe.

In the summer of 2016, Scharf's captain at the sheriff's department handed him a brochure about a cutting-edge DNA technology called Snapshot. A Reston, Virginia, company he had never heard of—Parabon NanoLabs—claimed Snapshot could use genetic profiles from crime-scene DNA to generate pictures of suspects. Parabon, in effect, was creating genetic mugshots. The brochure listed a number of cold cases around the country already solved by this new technology.

"I want you to write up a funding request to try out this Snapshot service," the captain said. "You pick the case."

Scharf chose Jay Cook and Tanya Van Cuylenborg.

17

Facing a Killer

Skepticism among cops and forensics specialists ran high at the idea of creating a physical description of a suspect out of a smear of blood or other bit of bodily fluid left behind by a criminal.

For twenty years, turning trace evidence into a DNA fingerprint had become the most treasured tool in the police arsenal for linking criminal to crime. The fingerprint analogy made sense of an otherwise incomprehensible scientific magic trick. But mining that same DNA to figure out the color of their eyes, hair, and complexion; the shape of their nose, ears, lips, and eyes; even whether they were prone to freckles or baldness or a unibrow? Generating a composite sketch of a suspect, something that human eyewitnesses often struggle to do? That sounded more like palm reading or fortune-telling to many cops, and Parabon's starting price of $3,600 per case for its most basic version of this service was high for many police departments on a tight budget.

Yet, when Parabon first introduced Snapshot in 2015, there were takers right away—detectives, chiefs, and sheriffs willing to try almost anything to solve the cold cases they just couldn't close. The skepticism vanished in a hurry with a couple of striking early results, starting with the French murders in Rockingham County, North Carolina.

Troy and LaDonna French were roused from bed on the night of February 4, 2012, by the screams of their nineteen-year-old daughter, Whitley. She had been awakened by an intruder in her darkened upstairs bedroom. Her assailant had tried to threaten her into silence with a knife, but she wouldn't cooperate and he fled the room, pounding down the stairs. The Frenches, meanwhile, had raced from their bedroom and were at the stairway landing, ready to go up to see what was wrong. The intruder exchanged his knife for the gun in his pocket, then shot and killed the couple as he fled—but not before he nicked himself with the knife and left drops of his blood on the banister.

Crime-scene technicians ran DNA from the blood drops through CODIS, but the search came up empty. The police then persuaded more than fifty men who had been in the French home or had some connection with the family to provide DNA samples voluntarily, including the daughter's boyfriend, John Alvarez. Traditional STRs for DNA fingerprinting were extracted from these samples and also a separate Y-chromosome profile, which shows whether two males share a paternal line—the same method used to prove Sally Hemings's male children and their descendants shared DNA with Thomas Jefferson. None of these comparisons yielded a match to the Frenches' killer.

Out of options, detectives decided to try Parabon's Snapshot in May 2015, a last-ditch effort. Expectations were low at the Rockingham County Sheriff's Office, but when the report came back, they were astonished. The description Parabon produced closely matched the boyfriend's brother, Jose Alvarez Jr. Eyes, ears, nose, and facial shape all were on the money, and the detectives thought the DNA mugshot strikingly similar, too.

But there was confusion. Jose Jr., along with the brothers' father,

had been ruled out already without even being tested, for an incontrovertible reason: brothers always have the same Y-DNA, passed down from the father. John's biological brother could not be the killer. Y-DNA matches are not always proof of guilt—many people, some only distantly related, can share a similar paternal line across generations. But an exclusion is solid and powerful evidence of innocence.

That was true, Parabon's scientists said, but cautioned that the detectives should not be so quick to assume that everyone's parentage is what they believe it to be. No matter what parents or birth certificates say, only a DNA test is definitive.

Sure enough, once a sample was taken, Jose Jr.'s DNA fingerprint was found to match the killer's. The man who raised him was not his biological father, a family secret that almost allowed a killer to elude

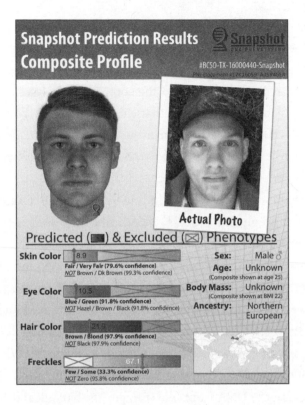

justice. Jose Alvarez Jr. eventually confessed and received two life sentences without possibility of parole.

In another unsolved murder, in Brown County, Texas, Parabon's DNA mugshot was so spot-on that the killer immediately went into hiding. Later, Ryan Riggs turned himself in and confessed to the sexual assault and fatal beating of Chantay Blankinship, a mentally challenged and childlike twenty-five-year-old who attended church with Riggs and his family. Riggs had never even been a suspect. He's now serving a life sentence.

These cases did not scientifically validate Parabon's work nor did all Snapshots have such dramatic results. Still, more than a hundred police departments came to Parabon in the next two years with cold cases. Some researchers and civil rights attorneys pointed out that the process hadn't been peer reviewed, but there have been no court challenges. There was nothing to challenge, because none of Parabon's findings were used to prove guilt nor were they introduced as evidence in a trial. The police simply used the company's analyses as a tip service to focus an investigation. If that led to a suspect, traditional (and thoroughly vetted) DNA fingerprinting technology would be used to confirm or rule out a match and to serve as evidence to support the filing of charges.

Jim Scharf was excited to get in line to have Jay and Tanya's killer portrayed through this process. It could be just what he needed to jump-start the case.

But then progress ground to a halt. First it took months to get approval from Snohomish County to move forward. Then Skagit County officials were brought into the discussion to see if they would pay half the cost. By the time that was settled, seven months had passed since Scharf's captain first handed him the Parabon brochure. When Scharf finally got the purchase approval, he encountered more infuriating delays retrieving the DNA from cold storage. The biological material was finally shipped off to Virginia, and once again it was time for Scharf to wait to hear back from a lab.

To SCHARF AND his cold case colleagues around the country, Parabon seemed to materialize out of nowhere. This was deceptive. The company's scientists and software engineers, led by CEO Steve Armentrout, a former CIA image analyst and artificial intelligence expert, had been working on this technology for many years—but in secret. Snapshot began as a Pentagon-funded anti-terrorism project.

Parabon had been founded originally as a grid computing business, creating and running software that harnessed many networked desktop computers to function together as a single, immensely pow-

erful supercomputer. Parabon's initial customers were in medical research, and the problem occupying most of their grid computing time
was DNA analysis. Medical researchers were looking for genetic clues
in the DNA molecule associated with a range of diseases and inherited traits—heart disease, cancer, Alzheimer's. The goal was to find
the parts of the genetic code responsible for such diseases and—this
was the holy grail—to develop ways to counter the effects of those
DNA segments, or simply turn them off at the molecular level. Turn
off the gene causing the disease and you turn off the disease itself,
at least in theory. Even the effects of aging on living cells could be
reengineered this way.

Sensing an opportunity in which there were far fewer players than
in the highly competitive grid computing space, the company spun off
a new division, Parabon NanoLabs, with two missions: to analyze and
"read" the actual content of DNA, and to build nanotech treatments
and vaccines with DNA, all using their own powerful grid computer
systems. The business sought to carve out a niche at the intersection of
biology and computer science.

The type of DNA structures that medical researchers were decoding
on their servers were SNPs, the same single nucleotide polymorphisms
that over-the-counter consumer DNA testing companies used. SNPs
were where the action was for serious research as well. Although they
represented only a small fraction of a human being's whole genome,
they still contained immense amounts of information on the actual
blueprint of a unique human being. Figuring out how to read that blueprint, to correlate specific clusters of SNPs on the DNA molecule with
specific traits—even basic ones, such as body type, ear and nose shapes,
eye color—was not a biology problem. It was a computational problem.
And Parabon had all the computer power needed to solve it.

When the company's early research showed promise, Armentrout
reached out to the forensic science community at a major conference in

2009 to suggest SNP decoding could be the future of DNA crime fighting—that in a few years, he could predict, for instance, what a person looked like in detail from their DNA. The reaction was immediate and adamant: "No you can't." They had their DNA fingerprints: proven, quick, affordable, and utterly reliable in identifying bad guys. Why mess around with expensive, exotic, and unproven SNPs? The idea, they said, was laughable. It presaged the response CeCe Moore would later receive when she pitched genetic genealogy as a crime-fighting tool at major forensics conferences. They just didn't get it.

Then along came the military, which was highly interested in what Parabon was doing. During the height of the Iraq and Afghanistan Wars, the Pentagon's shadowy Defense Threat Reduction Agency launched a grant competition by posing this question to a group of tech firms: If we discover and defuse an improvised explosive device before it blows up our troops, what can you tell us about the bomber from trace DNA left on the bomb?

Parabon won the competition and a grant under the federal Small Business Innovation Research program, and out of that grew the Snapshot technology and a related kinship prediction system. The company delivered a system that could analyze trace DNA on a bomb and predict what the bombmaker looked like, the region he came from, and what kinship he might share with other known terrorists. The new software could predict kinship of the bomber out to six degrees of separation—the equivalent of fourth cousins (relatives who share a great-great-great-grandparent). No other forensic test had ever achieved that, and it gave the military the capability to match people in custody to an unknown bomber.

To accomplish this, Parabon recruited more than ten thousand volunteers to provide a combination of SNP DNA profiles, 3D digital scans of facial features, and descriptions of body features. The grid computer power was harnessed to sort for associations between clusters of SNPs and physical traits, a process called "phenotyping," and it became the secret sauce of Parabon's Snapshot. The company also developed a capability to work with very small and very degraded DNA samples, such as might be found on a bomb casing in battlefield conditions—or at a civilian murder scene.

In effect, Parabon had built the Pentagon a supercharged 23andMe for terrorists.

Under the small-business grant program, once Parabon delivered its technology to the Pentagon, it was allowed—even encouraged—to privatize Snapshot and take it public. The forensic science community might be dubious, Armentrout predicted, but the cold case detectives were going to love it.

SCHARF GOT HIS report back from Parabon in March 2017. As he read the analysis and looked at the drawing of a man with sunken eyes who killed Jay and Tanya, Scharf felt both chilled and thrilled—and closer than he had ever been to an answer in the case. The summary read:

SEX: Male
SKIN COLOR: Fair or Very Fair with 90.4% confidence.
EYE COLOR: Green or Hazel with 93.3% confidence.
HAIR COLOR: Blond or Red with 80.4% confidence.
BALDING: Balded with age with 96% confidence.
FRECKLES: Few or Some with 25% confidence.
GENOMIC ANCESTRY: Northern European with 90.81% confidence.
FACIAL MORPHOLOGY: Larger and protruding forehead, brows and chin, wide cheeks and jaw, recessed eyes, narrower nose, and smaller top lip (compared to average predicted face for his gender and ancestry).

Parabon used this data to generate a somewhat generic, hairless baseline computer image of the killer's head and face. Then the company's forensic artist, Thom Shaw, "humanized" the 3D portrait by making substitutions from a vast catalog of eye shapes, nose types, hair qualities, and other features that matched the DNA predictions. He also "aged" the portrait to show the suspect's predicted appearance at various stages of baldness and longevity.

With all this in hand, Scharf started casting about for a way to put a name on that face. Yes, they would comb through all the old tips and home in on suspects who resembled this profile. Yes, they would hold a press conference and ask the public for help identifying this killer. But he wanted more than that.

First he called a genealogist who specialized in quick Y-DNA database searches, Colleen Fitzpatrick, and asked her to come up with names that might belong to the killer. She agreed, and for $950, Fitzpatrick produced a list of seven possible last names for the killer: Martin, Cole, Wolf, Gifford, Pelham, Amneus, and Correia-Neves. Scharf eagerly investigated, but the Y-DNA findings once again left only disappointment. There were a couple of crooks named Martin in the old files, singled out by tipsters over the years as possible suspects in the murders, but Scharf cleared them from the long list of persons of interest. None of the other names led anywhere. And with no one in old files who matched Parabon's DNA mugshot or profile, it was time to go public and see if there was anyone out there who recognized their killer.

DETECTIVE JIM SCHARF stood at a podium crawling with microphones, wearing his best black suit with a light blue shirt and tie. His bald scalp shone in the TV lights as he announced the new DNA evidence in the long investigation of Tanya and Jay's case. It was April 11, 2018. After months of frustrating delay as the two counties haggled over the where and when and how of the press conference, Scharf could finally show the world what Tanya and Jay's killer looked like—and to plead for help from anyone who might recognize him from these images.

"We believe someone out there knows something," Scharf said, his slow drawl still hinting, even after so many years in Washington, at the midwestern twang of his birth state of Illinois. "Help us solve this horrible crime."

He was flanked as he spoke by pictures of the murdered couple and maps showing their last trip on one side, and by three large posters of Parabon's DNA mugshot on the other. The three portraits depicted the killer's predicted appearance at age twenty-five, forty-five, and

sixty-five. The last and oldest, most likely closest to the killer's current appearance, was positioned prominently on its easel and seemed to glare at the audience.

No one could remember the last time there had been a big press conference called to discuss the murdered couple. Most of those heavily involved at the start of the case were retired. Some had passed away.

Several members of the Cook and Van Cuylenborg families had driven down from Vancouver Island to be there—the first time since the murders.

Jay's sister Laura Baanstra had still been in high school when her brother occupied the front page. Now she had kids of her own in high school. She had coped with the lack of answers all these years with a fervent certainty that whoever did this by now occupied a grave or a prison cell, for surely the person who killed Jay and Tanya had been caught for other terrible crimes. She refused to believe the culprit had roamed free all these years. Tanya's brother, John Van Cuylenborg, believed that was exactly what the killer had been doing, living the life he stole from his little sister. But John had, after many years of effort, focused on things he and his family actually could control, and he reined in the habit of wondering every day if *this* would be the day that answers came at last. His father had never been able to do that, and his son believed that it had destroyed him. Yet, in spite of that resolve, John Van Cuylenborg had allowed himself to hope again with this latest news in the hunt for his sister's killer. And so he had come.

"I had pretty much given up hope," Jay's dad, Gordon Cook, said before the press conference. He had not come. He was glad to learn that the case was being pursued with new energy and focus and a search for fresh tips. He was also afraid, for he had seen this surge of interest before. Three decades ago, he had spent the longest week of his life waiting for a different end to Jay's disappearance. There had been a flood of promising tips then. Another flood came in 1989, after the TV show *Unsolved Mysteries*. And again after news coverage about the Taunter being caught. Each time they hoped. Each time the hopes were dashed.

This time, though, they had pictures. And so Scharf painstakingly recapped Tanya and Jay's last journey, where they stopped, where they snacked and chatted, where they died. He also took pains to outline what was taken by the killer: backpack, jacket, camera.

"Maybe a relative who looks similar to one of these composites gave you a Minolta camera. Or you might have bought a camera like this around that time. The smallest detail could end up being the lead we need to solve this case. . . . All we need to do is get a sample of his DNA to be able to make the match."

Laura Baanstra walked stiffly to the microphones next, refusing to look at the posters depicting the killer's image. Later she would tell her husband, "I couldn't look at that face. I won't."

Staring straight ahead at the cameras, she made a plea in a clear voice with only the slightest quaver. "If these new pictures of this amazing new technology trigger a memory you had, perhaps of someone who said something odd . . . please, for the sake of my brother, Jay, Tanya, and all of our families, call it in."

And so they did. Once again, one hundred thirty people called the sheriff's hotline, offering new tips and possible suspects: men who looked like the computer sketches, men who bragged about killing a couple years ago, men with a Minolta camera back then, men the callers thought just seemed capable of doing something like that. Scharf knew better than to get too excited. Most of the information, almost all of it, would be well-intentioned but useless suspicions and gossip. Some of it would be a transparent and evidence-free ploy to get reward money, still being offered by Van Cuylenborg and Cook family and friends. There'd be the malicious revenge callers, perhaps a few delusional folks and habitual confessors. But mixed in the morass could be that one vital clue.

One caller—Tip #36—stirred Scharf's interest because it involved a possible sighting of the killer in the right place and time. In November 1987, this caller had been bartending at Snuffy's Tavern in Bellingham, across the street from Essie's, the bar where Tanya's wallet and ID cards and Jay's keys turned up. When he saw the Parabon mugshots on TV, the bartender told Scharf, he realized he had served someone who looked like the killer during the week of the murders. It had been a busy night at the bar, which led him to think it was the weekend, sometime between seven and eleven o'clock. He told Scharf the man looked like the Parabon DNA-based images, in his midtwenties, skinny with longer dark, dishwater blond hair that fell over his forehead and into his eyes. He talked with his hands and wore a green army coat. But the detail that most caught Scharf's attention was the traveler's checks. This customer cashed three Bank of Canada traveler's checks at the bar. This man had additional checks, but the bartender said he refused to cash any more. He couldn't remember the man's name, but he told Scharf he'd meet with him soon for a more detailed interview and help a police artist create a sketch of the man.

Scharf felt the tip could be solid, but that it might not really advance the case unless the bartender could remember something else that would help make an identification. He also felt highly suspicious of this tipster, who behaved oddly, knew a little too much, and talked about going to the Greyhound bus depot next to Essie's to go to Seattle for ballgames downtown—just a few blocks from where Jay and Tanya would have spent the night in their van. Scharf felt fairly certain that Jay and Tanya's killer had fled on a Greyhound bus after ditching the van, Tanya's wallet, and the other incriminating evidence at Essie's. The bus terminal was right there—it made sense. This bartender could just as easily be the killer himself, and Scharf asked another detective to see if he could surreptitiously obtain something with the bartender's DNA on it. He wouldn't be the first killer Scharf had encountered who deliberately called attention to himself by pretending to be a witness.

Then again, Scharf was suspicious of everyone. Most likely, none of it would lead anywhere, he grumbled. Either way, he would still have to sift through all the other tips, panning for gold in the mud. If only he could get the genetic genealogy investigation of the case going. It could help make sense of all this information, give him some names to match up with Parabon's Snapshot and the tips. Barbara Rae-Venter had called him two months before the press conference to say she had reversed her position on helping with Jay and Tanya's case. She felt she could do criminal cases after all. She just had to finish this big project she'd been working on and then she could get started. She'd also get cracking on Precious Jane Doe.

But Rae-Venter, too, had been delayed. It was April and still there had been no word from her, and it was one delay too many. The most patient detective in Snohomish County had finally grown . . . impatient.

This showed most often in the little things. It showed when irritated pacing replaced Jim Scharf's usual Zen calm while waiting for the arthritic elevators at the Snohomish County Courthouse to open their groaning doors. It showed in his unaccustomed fidgeting in line at Arby's while the server dithered over his buy-one-get-one-free roast beef sandwiches with Horsey Sauce. And it showed in the way he hammered his institutional-beige office telephone when people wouldn't return his calls, stabbing the gray buttons so hard it made the towers of papers on his desk quiver and shift.

It had reached the point where he was seriously considering putting in for his retirement. Maybe it was time to turn the cold case work over to someone else.

And then he heard from Barbara Rae-Venter again. She was ready to dig in on Jay and Tanya. They just need to get a SNP profile to analyze and upload for a genetic genealogy search, and she could get started.

At first, he thought he'd have to send out another round of samples for extraction. This posed a problem. The lab that had done such exceptional work with Precious Jane Doe had a huge backlog now. Scharf groaned: more delays.

Then he remembered that Parabon had already done a SNP profile a year ago for their Snapshot work. Perhaps, Scharf thought, he could save both time and money. He asked Rae-Venter if that existing profile would work for genetic genealogy, too, and she said it should. All the company had to do was email it to him.

Finally, Scharf thought, he was going to get some momentum back on the case. He picked up the phone and punched in the number for Parabon.

18

Cold Fusion

2017–2018
Reston, Virginia

Jim Scharf wasn't the only person interested in snaring genealogists as investigative partners. The computer scientists at Parabon were suddenly interested, too, though for very different reasons. This serendipity was about to make life interesting for Jim Scharf.

It began with a personal connection. Paula Armentrout, Parabon's cofounder and wife of the CEO, had an uncle who had served in the Korean War. Gene Gibson had been missing in action and presumed dead since 1952. In 2008, Paula joined other MIA family members in a Pentagon briefing on efforts to identify the remains of unknown soldiers recovered from the battlefield (there are more than six thousand from World War II alone). An army expert told the gathering that, although active-duty troops now have DNA fingerprints on file, there was no such easy way to find the name of unidentified casualties from earlier conflicts. A sudden realization jolted Paula: the military faced the same blind spot as the police. Both relied on the same genetic "bubble wrap"—the same STR DNA fingerprints—that Alec Jeffreys pioneered decades earlier.

Paula began frantically texting her husband. "Shouldn't they be using SNPs? They could identify my uncle that way. They could identify all of them."

This led to a conversation between military DNA experts and Parabon, and eventually to a contract to do kinship analysis to help identify the war dead. But to get the job done, Parabon needed to increase the power and reach of their kinship prediction software and for that they needed more data. To get it, they wanted to recruit hundreds of volunteers who had researched their extended family relationships—people who had built out their family trees to the fourth-cousin level and beyond. In short, they needed genealogists.

Local genealogy societies for hobbyists and professionals have long been a fixture in communities across the country, and many have regular public meetings, workshops, and fairs. Paula Armentrout found out when the next genealogy fair would be held near the company's Virginia offices, then showed up with a card table and a stack of fliers. She explained to attendees what Parabon was looking for: families with lots of known distant relationships on their family trees. Such data would allow Parabon to program new DNA-decoding algorithms, with the goal of bringing answers to grieving families and giving casualties of war their names back.

The fair attendees showed polite interest, but Armentrout was disappointed when it became clear no one wanted to get involved. Finally, a kindly woman with a thick southern drawl explained the problem to Armentrout.

"Oh, honey, no one here cares about DNA," she said. Old-school genealogists loved records and photos and newspaper archives, she explained. Not this new science stuff. "What you need to do is go to a *genetic* genealogy conference."

Few, if any, at Parabon were familiar with the term "genetic gene-alogy." But a quick internet search turned up some fascinating stories about amnesiacs and adopted kids and foundlings all discovering their identities and their extended families through genetic genealogy. And the name that kept coming up the most in these stories: CeCe Moore.

Paula Armentrout had just decided they needed to reach out to Moore when she got a phone call—*from* CeCe Moore. She was about to be freaked out by the bizarre coincidence, until she realized Moore hadn't asked for Armentrout specifically. Reception had just routed the call to the company spokesperson. Moore got right to the point with what both Armentrouts would discover was characteristic directness. Moore's contacts in the genealogy community had heard about Para-bon's inquiries and passed the information on to her.

"I thought I knew just about everything going on right now in the genetic genealogy space," Moore said, "but I don't know you. Who are you and what are you doing?"

Armentrout smiled. Then she told Moore exactly what they were doing and why.

"I love this project," Moore responded. "I'm going to post it to our group."

The folks at Parabon didn't realize then what that meant. She was going to put it out on her DNA Detectives Facebook group—which had swelled to more than one hundred thousand members by then. The next week, Parabon was inundated with genealogists contacting the company with offers to help their quest to identify the war dead—far more than they needed or could handle.

After that, Moore and both Armentrouts began discussing ways in which they could work together. Once Parabon understood what Moore could do with the consumer DNA databases, coupled with her skills at archival searches, they saw her take on genealogy as the perfect complement to their Snapshot service. They could work on a cold case and offer police departments a detailed physical description of the crook, predictive DNA mugshots, and a list of named potential suspects far more accurate than the old Y-DNA last name lists. Such a product would become a must-have for virtually every police agency in the country.

But there was a rub. When Moore worked a foundling case or an adoption or an unknown parentage investigation, she had living, breathing sources of DNA who could spit into a tube and swab a cheek. They had every legal and ethical right to search for matches on Ancestry or 23andMe or any other consumer DNA company without violating the terms of service—those intentionally impenetrable user agreements that are both ubiquitous and habitually unread by most people. Most of the consumer DNA companies did not allow uploads of DNA from sources other than their own mail-in kits. There was no way to get a forensic sample loaded into one of those databases. Furthermore, even if there was a way to fake one, it would violate the terms of service, and CeCe Moore said she would not do that. The debate among genealogists about working with law enforcement was a contentious one, and Moore said she had to be cautious.

There was another option, though. GEDmatch was an obscure public database for amateur genealogists, run out of a tiny Florida bungalow by a retired businessman with a couple of computer servers. GEDmatch was created so that anyone with a genetic profile created through any of the consumer DNA companies could upload it into an open, searchable database. No additional spitting into tubes required. The DNA companies weren't crazy about this, but these digital profiles belonged to their customers, so they had to allow it. GEDmatch was taking what had been proprietary information and making it public, so a user could then search for matches across all the consumer DNA platforms even though he or she only took one company's test.

When genealogists say they are "uploading DNA," they are referring to the code that maps each SNP in a person's profile—a computer file, not some biological sample. Generally referred to as a "raw DNA file," there's nothing magical about its form, which is a common, plain text file in the format of a primitive spreadsheet, readable by any word processing app. It's extraordinary only for its length, filled with line after line of alphanumeric codes. These codes are roughly analogous to coordinates on a map, except instead of showing the relative positioning of cities, towns, and landmarks, they describe the locations and pairings of the four nucleic acids that are the fundamental building blocks of the DNA molecule: adenine (A), thymine (T), guanine (G), and cytosine (C). The lines in this sort of raw DNA file look like this:

SNP#	Chromosome	Position	Allele 1	Allele 2
rs3131972	1	752721	A	G
rs141175086	1	780397	C	C
rs115093905	1	787173	T	G
rs11240777	1	798959	A	G

There is one line for every SNP. The extra-rich profiles Parabon generates from crime-scene samples can have more than seven hundred thousand lines. It is a tremendous amount of genetic information, though only a tiny fraction of what would be contained in a full DNA molecular profile. Even so, with that information being uploaded by each GEDmatch user, and a million users on board by 2018, GEDmatch was a powerhouse search tool, its impact far exceeding its humble origins. And the more people who uploaded their profiles, the more powerful it became.

It would be the perfect crime-solving tool, CeCe Moore told the Armentrouts. The problem was, GEDmatch had terms of service, too, with specific rules about which profiles could be uploaded. Crime-scene DNA was not one of them.

There was no actual enforcement of the terms of service, and when users violated them, there was no way to tell. If a cop (or anyone else) created a fake account on GEDmatch and uploaded a DNA profile of a killer he wanted to find, there would be no way for the owners to detect it and no enforcement mechanism if they did. GEDmatch operated on the honor system. And because the database was open to the public, there would be no constitutional violation of privacy in doing that. No search warrant would be required under current law.

But Moore explained to Parabon her own concerns: how she had been an evangelist for testing and for GEDmatch, and how she felt she would lose all credibility and betray those who came to her for help if she did not abide by the rules.

Moore's position on this was shared by many of her colleagues, though not all. Others thought it would be perfectly okay to use GED-match to catch killers. They pointed out that GEDmatch warned users in the terms of service that it couldn't prevent people from using the site for purposes other than pure genealogical research, including law-enforcement agencies investigating crimes. Anyone who still uploaded their DNA to GEDmatch was on notice—consent for police use was implied, this faction argued. Moore, though, felt there had to be more public discussion and awareness before making that leap.

But in their discussions, she and the Parabon staff realized there was something embedded in the GEDmatch terms of service that might open the door for them without controversy: You could upload the DNA profile of a dead person. That would include DNA from a deceased crime victim—a Jane Doe.

So in spring 2018, Parabon and CeCe Moore agreed they would start partnering on Doe cases, using genetic genealogy to identify murder victims and other anonymous dead. Success in that area could open up more possibilities for Parabon and Moore working together, perhaps leading to the sort of widespread discussion and knowledge about law-enforcement use of GEDmatch that would relieve any ethical constraints.

But all that changed a few weeks later. Moore's concerns became moot. Somebody else had let the genie out of the bottle.

WHEN JIM SCHARF reached Parabon and asked for the raw DNA file from his case, his sales contact at the company said he wasn't sure that was possible. He'd have to check with the company's legal department.

Most times, Scharf might have rolled his eyes and said fine and just waited for the red tape to run its course. But at this point, he had no patience for what smelled to him like bureaucratic BS. He was irked at being told that some house lawyer would have to sign off on emailing him his own forensic profile, paid for by his department. That file was his: it was evidence in his murder investigation.

Maybe, Scharf told the sales rep, he could resolve any legal concerns by serving Parabon with a search warrant for evidence in his murder investigation if they didn't get a move on. He was not joking. He added, "You think they'd respond to that?"

"Well, I guess they'd have to," his Parabon contact said. He promised that Steve Armentrout would call him right away to straighten things out. And Armentrout did call back, but Scharf was in a meeting by then and they missed each other. Still, Scharf felt certain he'd get what he wanted, and soon. Things were looking up.

Then the next day, April 25, 2018, news broke about a major cold case in California: police in Sacramento had arrested a man they claimed was the infamous Golden State Killer. Joseph James DeAngelo Jr., a seventy-two-year-old ex-cop, had been linked to at least thirteen murders, fifty rapes, and one hundred twenty burglaries in California committed between 1973 and 1986. This ultimate serial murder/serial rape cold case commanded huge news coverage. But, initially at least, key details remained unclear, including how they finally tracked down a cunning killer who had stayed one step ahead of the police for nearly a half century.

But Jim Scharf knew: this huge case had been cracked by genetic genealogy. Barbara Rae-Venter had phoned him that afternoon and told him she had been the genealogist working the case from behind the scenes, and she preferred to remain publicly anonymous. This was the case that kept her from jumping into Jay's and Tanya's murders sooner, but she said Scharf should also be reassured: catching the Golden State Killer showed she could do it. It had taken many months, with false starts and misplaced suspicions, and suspects to eliminate along the way with an entire team working on the genealogy, but they had gotten there at last. She was sure her methods would work for Scharf's case as well.

SOMEBODY ELSE GUESSED the Golden State Killer case was a genetic genealogy solve that first day: CeCe Moore. Someone had uploaded the killer's DNA to GEDmatch and reverse engineered his family tree—it was the only explanation, Moore surmised. Detectives had found another genetic genealogist to do what she had refused to do.

Moore was proven correct when press coverage the next day and in the coming weeks detailed the role genetic genealogy played in the Golden State Killer's arrest. GEDmatch was specifically named in most of the print and broadcast reports. The major news programs on every network were fascinated by the story, the combination of true crime and high tech irresistible.

Moore and Steve Armentrout began conference calling daily to discuss how this latest news might accelerate their dive into cold case investigations. They both felt such pervasive news coverage put the world on notice that law enforcement was using genetic genealogy and DNA databases such as GEDmatch. That knowledge meant people could choose whether they wanted to keep their profiles on the site or remove them. When GEDmatch posted a public statement on the website two days after the Golden State Killer arrest, any ethical concerns Moore and Parabon harbored had evaporated:

> We understand that the GEDmatch database was used to help identify the Golden State Killer. Although we were not approached by law enforcement or anyone else about this case or about the DNA, it has always been GEDmatch's policy to inform users that the database could be used for other uses.... While the database was created for genealogical research, it is important that GEDmatch participants understand the possible uses of their DNA, including identification of relatives that have committed crimes or were victims of crimes. If you are concerned about non-genealogical uses of your DNA, you should not upload your DNA to the database and/or you should remove DNA that has already been uploaded.

It didn't get any clearer than that, Moore told Armentrout. Anyone who objected to the police using their DNA to solve crimes was on notice and had every opportunity to pull their profiles from public view. Moore was all in: "I'm ready today."

Armentrout was ecstatic. He told Moore that Parabon had over a hundred DNA data files of cold case crime-scene DNA that had been used for Snapshot phenotyping—cases that were still unsolved. He could upload them all to GEDmatch in a matter of minutes and she could get to work on the genetic genealogy as soon as the police agencies gave permission.

And the first one he wanted to tackle, he told Moore, was the case of Tanya Van Cuylenborg and Jay Cook. That case would be their proof of concept.

WHEN HE AND Scharf spoke on the phone the next day, and the detective asked once again for the DNA profile to give to Barbara Rae-Venter, Armentrout said he had an alternative to suggest.

"I think we can help you out with that. How about before you hand it off to someone else, we work up a preliminary genetic genealogy report for you instead? We'll do it for you pro bono."

"Well, I can't turn down that offer," Scharf said, laughing, and the deal was struck, the talk of search warrants forgotten.

Armentrout said he'd have Scharf's profile uploaded to GEDmatch right away and get him a report that would include the number and degree of usable matches they found, and whether additional genealogical research would likely lead to the killer's identity. Scharf figured that meant he'd get a list of some thirty or more possible suspects, and then there would be an arduous investigation ahead. Based on what he had read about the Golden State Killer and what Barbara Rae-Venter told him, it had taken months to sift through the matches and puzzle out the genealogy. He expected something similarly difficult with Tanya and Jay.

"By next week," Armentrout said as they ended the call, "I should be able to tell you who your suspect is."

It was Thursday, April 26. Next week was only four days away.

"Yeah, right." Scharf chuckled. If only.

19

He's the One That You Want

As excited as he was at the potential for a break in the case, Jim Scharf was not about to stop pursuing other promising leads. If Armentrout could deliver on his big talk, great. If not, well, Scharf would keep the case moving forward. Plus, he admitted to Laura, he had to keep busy or the waiting would drive him crazy.

So he continued researching a tip from the early investigation that hadn't been checked out properly back in the eighties—a witness who said he saw the Cooks' copper-colored van gassing up at the Bryant General Store in Arlington. The tipster thought he recognized the man at the pumps as a former coworker he hadn't seen in fourteen years, Terry Nichols. Scharf found a photo of Nichols and thought him a good fit for the Parabon Snapshot description. The Bryant General Store was only a half hour's drive from where Tanya's body was found, and it was a plausible gas station stop for someone taking the back-road route after leaving Jay Cook's body at High Bridge. Nichols had died years ago, so Scharf was trying to find a relative willing to submit to an ancestry DNA test to either clear Nichols or tag him as a viable suspect. He even contacted a former mayor of Seattle with the same last name. The retired politician said he was happy to help but sorry to disappoint. He and his family had done their own DNA ancestry tests a few years earlier, and he assured Scharf that his suspect Terry Nichols was not his relation.

Scharf lined up the department's sketch artist for the next day to follow up on his other recent lead. He and the artist spent Friday driving up to Bellingham together to meet with the bartender from Snuffy's Tavern. Scharf wanted a traditional composite drawing of the man who had been passing Canadian traveler's checks a block away from Jay's van and Tanya's discarded wallet. Maybe this would be the face of one of the people on the GEDmatch list he was about to receive. Scharf felt he was making more progress on the case now than at any time since he had hunted down the taunting letter writer eight years before.

While the witness was occupied with the artist, Scharf checked his email and was irritated that there was nothing from Parabon. He fired off an email to Steve Armentrout: "Did you upload the profile to GEDmatch like you said you would?" The response came quickly: "It's in the works." The detective was mollified. Mostly.

But Scharf had no idea what "in the works" really meant. If he knew, he would have been doing handsprings. Armentrout hadn't been joking about naming the suspect by next week. In fact, though he didn't know it yet, he was underselling what was about to unfold. The GEDmatch list of possible relatives came back late in the day Friday. Then it was CeCe Moore's turn. By Saturday morning, already ensconced on her beige couch with coffee and laptop, she sensed that she would break the case, and quickly.

There was a bit of destiny involved with this, Moore thought. She wanted this one to be her first cold case. It's the one she would have picked if asked. She had read the press coverage and immediately felt drawn to Tanya Van Cuylenborg—and the eerie parallels in their lives. Both she and Tanya were the same age; they graduated from high school the same year. Moore's father grew up in Canada near Tanya's home, on tiny Gabriola Island, just a mile off the coast of the much larger Vancouver Island. The whole area was Moore's ancestral homeland, and growing up, she spent many summers on Vancouver Island and in Seattle. She rode the same ferry between Victoria and Port Angeles that Jay and Tanya took more times than she could count. She felt very connected to this case and to Tanya, and she dug in with a relentless determination to find out who killed the young woman who so easily could have been her.

Moore's search began with the two most promising names on the list from GEDmatch: a man and a woman who possessed gene markers that pegged them as second-cousin-level relatives of the killer. Second cousins have an average of 3.13 percent of DNA SNPs in common. The average person might think such a percentage made for distant cousins, but for a genealogist, it was almost like they were in the same room. Often the closest matches Moore finds are third and fourth cousins, which means having to build family trees that go many generations back in time to find the common ancestors, a very wide and time-consuming net to cast. Second cousins, though, were gold.

Even better for Moore, the two cousins at the top of the match list, though related to the killer, were not related to each other. They were in different genetic networks, meaning one cousin came from the killer's father's side of the family, while the other was related to the killer's mother. One was a Talbott, the other was a Rustad. Somewhere across the generations those two families connected through a marriage, and the family-tree branch that grew from that union included a killer. As soon as Moore saw that, she was hooked, the kid with a sweet tooth peering in the candy store window. She promised herself she would not sleep or stop until she had the goods.

The next step meant digging into archive searches of newspapers, funeral home obituary sites, and general social media stalking. She spotted plenty of Talbotts and Rustads online, but nothing that connected the two families. She moved on to obituary archives and found a 2015 obit for a seventy-three-year-old woman named Patricia Talbott. The woman's maiden name had been Peters. A birth records search told Moore that Patricia's mother was named Blanche Peters. And the next set of records held the key: Patricia's mother was born Blanche Rustad.

With that, Moore knew that Patricia Peters's marriage created the genetic crossroads that would lead to a killer. Her union with a man named Talbott provided the one and only link between the Rustad and Talbott clans. If that couple had children, they (and their offspring) would be the only people on the planet who shared DNA with both the Rustads and the Talbotts. The killer had to be one of them.

She sifted through the family's Facebook posts and found them to be a boisterous, loving, warm group. Patricia had been a kind and quirky matriarch who coached soccer, made jewelry, gardened, and recycled old fur coats she found in thrift shops into stuffed bears for children. Patricia Peters had married William Earl Talbott Sr., and they had been together for fifty-five years when she died of kidney failure in 2015. The couple had raised their children and lived most of their adult lives in the Seattle area, not far from the Snohomish County line.

Patricia and William had four kids, three of whom were active on social media, sharing pictures and stories and jokes, while one of them was a ghost. He had no footprint, no photos, no presence, Moore found.

The killer left semen behind, so the suspect had to be male. If there had been four brothers, Moore would have given four names to the police for them to investigate. But there were three sisters and one brother—the sibling who was the social media ghost.

It had taken at least two months of genealogy research to put a name and a face on the Golden State Killer. Moore identified Jay and Tanya's likely killer in two hours. She would spend the rest of the weekend double-checking her findings, just to make sure. But at ten thirty Sunday night, she took her hands off her laptop keyboard and let herself exult, thinking, *I know something no one else in the world knows.* She savored that for all of thirty seconds. Then she started typing again—a short email to both Steve and Paula Armentrout. The subject line had one word: "Solved."

The message had six words: "You might want to call me."

On the other side of the country, the Armentrouts read that message and had a classic "holy cow" moment. "She did it," Steven Armentrout said.

Then they both had the same thought: When can she start on the next case? As they saw it, this wasn't going to be a one-off. This was going to be a cold case revolution.

THE NEXT DAY, a Monday, Armentrout emailed Jim Scharf and asked him to call. It was the detective's day off and he was out walking his two aged pugs, Harley and Cheyenne, when he reached the Parabon CEO.

"I've got a name for you," Armentrout said.

At first the detective laughed, juggling his phone and his dogs, still thinking this was an exaggeration or a joke. He'd be happy to get a list of a dozen names to check out. Then, as he listened to Armentrout's recap of CeCe Moore's findings, the detective realized his list really did have just one name on it. Moore even came up with the address in Woodinville, Washington, where the suspect's family lived at the time of the murders. Scharf knew that location well from his days as a patrol deputy assigned to that part of Snohomish County bordering on Woodinville. The house stood—literally—just up the road from where Jay Cook's body had been left beneath High Bridge. The detective whooped, startling the pugs, who stared up at him in indignation. Then Jim Scharf felt the tears come.

After a few moments, the detective and the CEO agreed there'd be a longer conference call and briefing tomorrow morning with Parabon, the sheriff's department, and CeCe Moore. Then Scharf ran into the house with his pugs jogging along, yipping in confused excitement as he told his wife the news.

Scharf had finally decided he would join Laura in retirement at the end of the year, after more than forty years on the job. Now all he could think about was how this new method of genetic genealogy could be applied to his stack of other cold cases. He, too, sensed a revolution in the making. As he hurriedly changed into work clothes and texted his boss, he said to Laura, "I guess I'm not going to be retiring anytime soon."

At last he had a suspect, with a name and a face and a target on his back: William Earl Talbott II.

After thirty-one years of frustrated justice for Tanya and Jay, this cold case, at last, had turned red-hot.

PART III

MYSTERY MAN

But now he's alone and he sulks to himself
As he pictures the Mime dead still on the shelf.
This saddens the boy for now he has grown
And the Mime went and left him here on his own.

—"Time," by Tanya Van Cuylenborg

This is just beyond my comprehension,
being charged with something of this nature.

—William Earl Talbott II

20

———

Watching and Waiting

It was day two in this new stage of the Tanya and Jay investigation. The previous stage—until April 30, 2018, the *only* stage—was all about the mystery of who did it. That stage of casting a wide net and coming up empty lasted thirty years, five months, and seven days—11,116 days total—since the damp November morning when a recycling scavenger found Tanya's body on Parson Creek Road. Now Scharf was in stage two: the catch was in the net and the suspect needed to be pulled in with evidence of his guilt—or let go if the evidence wasn't there. Scharf's gut told him this would not be a catch and release.

Scharf arrived at the office with plenty of time to spare before the conference call with CeCe Moore. He went to his desk to look through the meager but growing file he had begun to assemble on William Earl Talbott II. He had looked up the name as soon as Armentrout gave him the preliminary information.

Curiously, there were two William Earl Talbott IIs in the Seattle area, one born in 1939, the other in 1963, father and son. Somehow some of the official records had made them both a "second," which couldn't be right but was confusing everyone. For a crime more than three decades in the past, either Talbott could be a viable suspect, one forty-eight years old at the time of the crime, the other twenty-four. From the preliminary information he had in hand, he couldn't tell which was his target.

Scharf's local genealogist and researcher, Deb Stone, did a quick background check of the family and reported that the elder Talbott went to school in Seattle, worked for *The Seattle Times* newspaper for forty years, and served on the city council of the very small and picturesque town of Cathlamet in southern Washington, where he and his wife moved after selling their house in Woodinville. The father was squeaky clean in terms of any sort of run-ins with the law.

The son, however, had mention of a single assault in court records in King County (where Seattle is located) dating back to October 1984, when he would have been twenty-one years old. Scharf had no details yet, but whatever happened, it had been treated as a minor offense, punished with a fine of $150 and a sentence deferred for one year, which usually means that it was a misdemeanor and, if he behaved himself for that year, the case would simply go away. So, technically, he had no criminal record and no other arrests. But the mere existence of the charge interested the detective.

The elder Talbott lived in the town of Kelso, the self-described "City of Friendly People" a couple hours south of Seattle. Scharf's research on the younger Talbott suggested he used two addresses interchangeably for official purposes. One was vacant land occupied by trees and a gravel road in the tiny rural town of Riverside, closer to the eastern edge of Washington than the coast. The other was a home on the western side of the state, in the city of SeaTac, best known as the town that abuts, and was named for, the sprawling Seattle-Tacoma International Airport. He had a four-year-old bronze-colored Jeep Cherokee registered at the Riverside address, and his latest driver's license gave Riverside as his residence. He was also registered to vote there, though he hadn't voted since 2011.

Since he was currently employed by a Seattle trucking company that made local deliveries, and he would have to endure a daily commute of more than four hours each way from Riverside, Scharf figured it was a safe bet that he had lived in SeaTac for years. Had he been trying to live off the grid at one point? Or was he making himself hard to find, giving the outward appearance to anyone trying to hunt him down that he lived on a vacant lot in another part of the state? Was he just a private person who liked to keep his true address hidden, or was there a more sinister reason for the obscurity? Even more interesting to Scharf, however, was the list of addresses he had given in the 1980s, which showed both William Talbotts had lived in the family home in Woodinville, not far from where Jay Cook's body was found at High Bridge. The area was secluded, and the stretch underneath the bridge was extremely private, invisible from the vantage of drivers on and approaching the bridge. Few but locals would know about that spot as a safe place for kidnapping, rape, and murder with little chance of outsiders stumbling into the scene. Excitement swept away Scharf's chronic tiredness.

He found the conference call with CeCe Moore revelatory. She walked Scharf and his colleagues through her family-tree triangulation using the two cousins from different family lines on the GED-match list. Then she explained how only one marriage brought those disparate family trees together: the elder William Earl Talbott to Patricia Peters. The father was not a "second," and records that claimed otherwise were wrong, she said. She assured Scharf that the elder Talbott could not be the killer, as he could not carry genes belonging to a cousin on his wife's side of the family. The person who left his DNA in the van and on Tanya had to be a male child of William and Patricia to achieve the mix of DNA of the suspected killer. There were four kids born in that household, but only one was male. The DNA never lied: the known genealogy said the killer had to be William Earl Talbott II, born on March 4, 1963.

Moore also ran down Talbott's siblings: his youngest sister, Malena, who lived in Oregon; sister Inga, who lived in Woodinville and was a year younger than Talbott; and his older sister, Siglindie "Angel" Talbott Fisher, who had lived for years with her husband and three children next door to Talbott in SeaTac. From social media posts, Moore said, it seemed Talbott had cut off all communication with his siblings and father. His mother died in 2015, but it appeared that he cut her off as well.

This keenly interested Scharf, who hungered to know and understand his quarry. Had Talbott retreated into hermitage after the murders? If he exiled himself at the end of 1987, Scharf felt, that could be powerful evidence of a killer's state of mind.

The detective wanted to talk to them all, but he knew that would have to wait. And then he learned he would never talk to the eldest sister, Angel. The warehouse worker, gardener, demolition derby aficionado, wife, and mother had just died of heart failure—on April 30, the same day Scharf learned Talbott was the killer. Angel, at least, would be spared what the rest of the family would soon experience: the horror of seeing one of their own accused of rape and murder. Scharf felt bad for them, but a part of him couldn't help but think, If only I could have gotten the genetic genealogy going sooner, I could have talked to Angel, the one most likely to know William Earl Talbott II best.

The first task, though, before talking to anyone who might alert Talbott that he had been exposed, was to verify that he really was the man who left his DNA at the scene of the crime. CeCe Moore expressed confidence in the accuracy of her genetic sleuthing and, after that conference call, Scharf felt the same. Her work seemed to him beyond reproach. But Moore herself pointed out the inherent limitation of working with DNA from kin instead of a suspect: Talbott's actual DNA was not in GEDmatch. Given his invisibility on social media, his withdrawal from family life, the measures he took to conceal his true home address, and his seemingly solitary existence, there is no way such a person would send his DNA to a consumer genetics company, much less upload it to a publicly shared database. Nor did Moore find any trace of his immediate family's DNA in there, either— just the two second-cousin-level matches, neither of whom were close enough to know Talbott in the least. All the connections beyond those cousins were based on birth, death, and marriage records, obituaries, and other documentary evidence.

For building family trees, that was plenty. For charging a man with murder and depriving him of his liberty, it was not enough. The advent of genetic genealogy, Moore explained, has revealed that, far more often than anyone realized, biological parentage is not always what families believe it to be or what archival records show. From marital affairs to secret pregnancies to adoption, there are any number of events that could have occurred a half century ago that would cause this identification to be wrong. Yes, a male child of Patricia and the elder Bill Talbott was the killer. But it's possible the person known today as William Earl Talbott II was not that child. The only way to know for sure would be to get his DNA directly from the source and see if it matched the killer of Tanya and Jay. Then, and only then, could the police feel confident in arresting him.

Scharf understood and appreciated Moore's words of caution and knew she was right. Parabon's DNA mugshot and description were very close to the real Talbott, he thought, but not perfectly so. The main miss in their report was the predicted eye color. Talbott has piercing blue eyes, whereas Parabon reported that ninety-three out of a hundred people with his DNA would have green or hazel eyes, while only seven out of a hundred would have blue.

Getting a sample from Talbott for a classic DNA fingerprinting without tipping him off was the challenge. The best way would be to scoop up something he discarded. Scharf has done this more times than he could count. Long-standing case law in Washington State allows the police to take a person's trash without a warrant, regardless of where it is left—a trash can, a restaurant table, a park bench, a city street. As long as it is abandoned, and there are no privacy rights attached, no search warrant is required.

To get that discarded item with Talbott's DNA, the sheriff's department higher-ups approved a surveillance and stakeout operation using their Violent Offender Task Force. While that was being organized, Scharf and his sergeant drove to SeaTac to check out where Talbott lived. Then they drove by his workplace, Nelson Trucking, and finally out to Woodinville-Duvall Road to see Talbott's family home, where he had grown up. Scharf checked the odometer and drove from that address to High Bridge. To get to the spot where Jay was found below the bridge, you had to go north and turn right at the first stop sign. Then the bridge was just ahead. As Scharf measured it, there were just seven miles between the murder scene and the house where Talbott and his parents once lived.

By the time Scharf got back to the office, the surveillance team had reported in: they had already started following Talbott. Scharf thumbed through a report from the state motor vehicle department that held every driver's license Talbott had received since the year 2000, with enlarged photos included. He had been hoping for something older, closer to 1987, but photos of Talbott were hard to come by. There was one blurry photo from someone's wedding, 1986 or 1987, before the murders. Talbott was twenty-three or twenty-four, wearing

a bright white tuxedo shirt and white tie, his youthful round face looking ruddy against the starched collar. He had a full head of brown,
bushy hair and a full, if scraggly, dark beard. He stared up from the
photo at Scharf with deep-set eyes, a slight, forced smile on his small
mouth. Then the detective studied the DMV pictures. There was a very
clear and dramatic progression.

The license issued in March 2000 shows a broadly smiling Bill
Talbott, white teeth flashing, and that unforced, genuine smile trans-
forms him in a way seldom seen on anyone's driver's license. He looks
pleasant, affable, friendly. He is thirty-seven in this one, now bald on
top, short-cropped hair on the sides, still dark. Compared to the wedding photo, his face is broader, heavier, clean-shaven except for a small
wispy mustache. His height is listed on the license as five ten, and he
has gained quite a bit of bulk since 1987, with a listed weight of 190
pounds.

Three and a half years later, the beard was back, he had less hair on
top, his smile was gone, and he had gained forty pounds, with a listed
weight of 230. His address, previously in Woodinville, was reported to
be his vacant lot in Riverside.

In 2009, at age forty-six, he was completely bald, the side hair
shaved, the facial hair trimmed down into more of a chin beard with
a mustache growing down into it, showing strands of gray. His license
was now the commercial permit of a professional truck driver. His face
was broader, pear-shaped with more weight, which was now listed as
240 pounds. The extra pounds made his eyes look smaller. His expression is . . . expressionless.

The most recent license was issued in 2014. Talbott was fifty-one.
His weight had risen to 260 pounds. His beard was salt and pepper now,
and it couldn't hide his double chin. His eyes looked sunken, dark
circles beneath. There was an unhealthy cast to his face, as if he was
getting over an illness when the photo was taken. He was showing that

same small, forced smile he wore in the wedding photo twenty-seven years before.

Scharf sat back in his desk chair. Judging by the photos and the reported weight, Talbott was a big man, overweight for his height, almost certainly obese. But he had been a trucker for years and was likely muscular and strong. He could be a formidable person to arrest. If it came to that—when it came to that—Scharf wanted to have a solid arrest plan and plenty of backup.

He recalled that incident from his patrol deputy days, his life-and-death struggle with a bulky and powerful manic-depressive who had nearly choked him to death. He'd never forget Dino Scarsella's expressionless eyes, massive forearms, and blunt fingers with their iron grip around his neck. He, too, had weighed 260 pounds.

The bulk and build of William Earl Talbott II—and the eerie coincidence of Talbott having exactly the same weight as Dino Scarsella—brought those old memories to mind. When Scharf had left patrol work to become a detective, there were far fewer occasions for ad-libbed life-and-death encounters and many opportunities for carefully planned arrests. He promised himself that this impending arrest of another big and powerful man would be more choreographed than a Broadway musical.

While the surveillance team watched and waited for the DNA to drop, Scharf had time to script everything—from how he would initially approach his unsuspecting suspect to devising some kind of ruse to get him in for a chat without formal arrest and reading of rights. He planned the actual arrest moment. He wrote search warrant applications. He devised questions for Talbott and his family. And he designed a press conference that would not just inform the public of a momentous cold case solve but would also bring new witnesses out of the woodwork. And all the while, he kept checking in with the surveillance operation.

Meanwhile, the old partnership with Skagit County, where Tanya's body had been found, was revived. Scharf had to get prosecutors in both counties on board so they could figure out who would do what and where charges would be filed. The head of detectives for the Skagit County Sheriff, Sergeant Jenny Sheahan-Lee, researched the movement of Talbott's work truck and found out it had crossed into Canada, apparently for one or more deliveries. A Talbott-Canadian connection, depending on how far back in time it went, could be a valuable new line of investigation.

Sheahan-Lee had a special connection to this case, she told Scharf. In 1987, when she was an eighteen-year-old with the sheriff's Explorer volunteers, she had participated in the search of the woods at the Parson Creek Road murder scene. She was the one who found the virtual needle in the haystack: the shell casing that matched the caliber of the bullet that killed Tanya Van Cuylenborg. Being involved in the case again, with a real suspect in sight, was deeply satisfying for her. Scharf took it as a good omen.

On day three of the surveillance, the team told Scharf they hadn't seen Talbott all day, and he hadn't gone to work. His Jeep and an old Dodge registered to him were parked at the house in SeaTac, but when one of the agents knocked on the front door, no one answered. Scharf thought he might have asked for bereavement leave from work to mourn the sister he had shunned for years.

Day after day, the surveillance team reported rarely seeing Talbott, if at all. He definitely stopped going to work. Had Talbott somehow caught wind that he was under suspicion? Had he spotted his tail? The team didn't think so, but his suspect's sudden invisibility made Scharf nervous.

Scharf next put in requests for Talbott to be run through several serial offender databases to see if his address, workplaces, or any other information had a connection to any known serial rapists or killers.

ViCAP, the FBI's Violent Criminal Apprehension Program; Washington attorney general's HITS, the Homicide Investigation Tracking System; and Canada's similar database were all consulted. Like his predecessors who worked the case, Scharf still believed Tanya and Jay's killer could have other victims. He even asked that the old files of the Green River Killer task force be combed for references to Talbott in connection with that most prolific of Washington's serial killers, who confessed to seventy-one murders in the 1980s and 1990s before he was finally identified as Gary Ridgway and arrested in 2001. Hundreds of suspected Green River Killers had been investigated and thousands of tips had come in along the way. But Scharf was told none of them implicated Bill Talbott.

Scharf also pulled one of his own cold case files, the five of clubs in his cold case deck, the murder of Sun Nyo "Julie" Lee, who he thought might have been another Talbott victim. The thirty-six-year-old's skull had been found near High Bridge in 1991, eight months after her boyfriend reported her missing. The rest of her body was never found. She had been bludgeoned, then decapitated. But other than the proximity to Talbott's family home and to where Jay Cook's body was found, Scharf could find no connection to Talbott—or anyone else.

But other things started to fall into place. Scharf reviewed Talbott's employment records and found he worked for an aerospace firm in the Seattle suburb of Kirkland in 1987, Hirschler Manufacturing. But he was let go in the fourth quarter of 1987, leaving him unemployed in November of that year, when Jay and Tanya were killed. That meant he had the free time to hunt down victims on a weeknight, kidnap and kill them, and leave the van and evidence in Bellingham days later. It also meant Talbott had no potential work-related alibi. There would be no time cards or pay stubs or coworkers to suggest he had been on the job during key events in the timeline of Tanya's and Jay's kidnapping and murders.

Next, Scharf went through all the evidence and photos gathered at the crime scenes. He verified that the flex ties found with each body and in the van were all of the same type. Then he realized that, in all the years since the murders, those plastic ties had never been examined for any DNA on them. He checked them out of evidence and brought them to the crime lab. If Talbott's DNA showed up on any of the ties, it would cement them as relevant to the case and not just random pieces of litter present by coincidence.

Finally, tips kept coming in about other suspects in response to the press conference about Parabon's DNA mugshots. These tips all suggested suspects other than Talbott, but Scharf dutifully checked them out as if the case was unsolved, which was officially still true. Until they had Talbott's DNA and matched it to the crime, it remained possible someone else had committed those murders. He also knew that if Talbott was charged and faced trial, his lawyers would inevitably check to see if investigators had overlooked or failed to follow up other promising leads in the case. If they could, they would use that fact to undermine the case against Talbott as slipshod or biased or blinded by tunnel vision. At this point, Scharf's investigation had to play defense as well as offense.

On Tuesday, May 8, Scharf's surveillance team reported that Talbott had finally resurfaced, left his house, and had gone back to work. The undercover cops resumed their tail, following him to the trucking company, then trailing behind as he made his deliveries with his big rig. Scharf tried to feel encouraged by this, but he wondered how many more days it would take to get the evidence he needed.

Not much longer at all, it turned out. That same afternoon, Scharf got the call from the surveillance team leader: Talbott had dropped that used paper coffee cup. That particular wait, at least, was over.

Taking charge of that piece of evidence, holding the incongruously vital bit of trash, and taking it to the lab felt more nerve-racking than exciting. What was it about this case that seemed to strip him of his career cop's armor of cynicism? Scharf wasn't sure. Maybe it was the idea that, if this one case went well, it would help revolutionize all his cold case work—maybe everyone's cold case work, proving that the Golden State Killer wasn't a one-time fluke. Or maybe it was how Tanya's and Jay's family members had reacted when Scharf had called to let them know there had been progress in the case at last, that he was investigating a promising new suspect. He could hear in their voices and cautiously worded responses how hope battled with their need to shield themselves from yet more disappointment and despair. Scharf had transported important pieces of evidence more times than he could count over the years, but rarely had so much rested on something so little.

The next day, the undercover team followed the trucker to a diner for his lunch break. When he sat down to eat rather than get takeout, one of the undercover officers strolled in and quietly recruited a waitress to scoop up and save Talbott's abandoned soda cup, paper plate, and plastic fork, just in case the coffee cup ended up not having enough usable DNA for the lab to analyze. That afternoon, Scharf drove back to the state crime lab yet again, bringing this new evidence to Lisa Collins, the DNA lab supervisor in Marysville. When he tried to give it to her, she told him to wait a bit and walked back out.

He didn't like sitting still. Keeping busy was the only way he could stand the waiting. But only a few minutes passed before Collins walked back in and said they wouldn't need to test that new fork, plate, and soda cup. Then she handed him the report on the coffee cup. He stared at the first page, scanning for the summary of the findings. There it was: the profiles matched. There was no doubt: William Talbott's DNA was present at the murder scenes.

He looked up to see Collins's grinning face. Then he started making the calls. He phoned his supervisors in Snohomish County, the detectives in Skagit County, prosecutors in both counties, and the sheriff himself. Finally, he called Tanya's brother and Jay's brother-in-law with the news: the murders of Tanya Van Cuylenborg and Jay Cook had been solved.

IT TOOK A week to put everything in place for a completely choreographed arrest. Scharf's plan called for teams of detectives from both Skagit and Snohomish to be positioned before the arrest, ready to interview Talbott's relatives as soon as he was in custody. The goal was to avoid even the possibility of them talking to one another before the police got their statements.

Scharf put together a slide presentation on the murders and the new suspect, photographs of Talbott and his relatives, and a list of more than seventy questions he wanted the detective teams to ask each family member. They ranged from whom he dated and whom his friends were, past and present, to whether he liked to hitchhike, ride ferries, or take Greyhound buses. He even wanted to ask if Talbott might have a twin brother (because twins have the same DNA).

Scharf also had a plan for how he would confront and, hopefully, interview Talbott at the trucking company. And he had an elaborate script for how he wanted his conversation and questioning with Talbott to go.

When everything was ready, the arrest day was set: Thursday, May 17, 2018.

21

Can You Come Back Tomorrow?

"There's a truck coming," the outermost member of the surveillance detail radioed amid a burst of static.

Inside his car, Scharf tensed. Was it showtime at last? The stakeout at Nelson Trucking was dragging out all day as they waited for their suspect to return with a rig emptied of deliveries. Scharf felt irritated, his patience again tested. He had wanted the team to begin tailing Talbott first thing in the morning, before the big man took his truck out. Instead, they had assembled at the sprawling Nelson depot on the south end of Seattle after all the trucks went out. Now they had to wait until the end of the day to finally make the arrest. Scharf reminded himself that he had been working toward this moment for the last thirteen years. The families of Tanya and Jay had waited for thirty-one. What were a few more hours, give or take?

"He's turning in," another cop's voice said. And then Scharf could sense the rumble and growl of the diesel; see its approach, the sooty fumes spewing from its chrome exhaust stacks; and hear the rhythmic *clang* and *thump* of the empty flatbed behind the tractor rattling its jarring metallic beat with every bump in the road. Scharf could feel the engine sound in his chest, bass drums in a parade.

"It's not him," another voice on the radio said, flat with disappointment. Everyone stood down as the truck thundered through the gate and drove downhill into the depot.

Five hours into this very high-stakes stakeout, with the suspect nowhere to be seen and the dinner hour fast approaching, the watch team members had begun to grow restless, hungry, and bored.

This is common in long, complicated, static surveillance operations, even momentous ones such as this. Tailing was different, active, a constant challenge to maintain line of sight without becoming noticeable. Sitting still for hours was harder. Some of the cops on the detail passed the time by shooting the bull with their partners or talking on their cells. Others moved their cars out of hiding and pulled next to colleagues, allowing them to chat privately through rolled-down windows.

Scharf did none of those things. He spent those five hours in his car obsessing over his arrest plan. His pale blue eyes scanned the vast grounds of the depot, big enough to accommodate four football fields. Scharf looked for avenues of escape for Talbott if he decided to flee. As vehicles came and went, the detective considered ways that the backup team could move in covertly, hiding behind vehicles to avoid tipping off Talbott to the police presence. Scharf mentally rehearsed the script of affable, low-key lines he wrote for himself to persuade the suspect to talk rather than clam up and demand a lawyer. That's why the backup needed to lay low.

The arrest plan had required lengthy negotiations between the two counties. Legally, the charges could have been filed in either, but Skagit wanted it there, arguing that Tanya's was the stronger case, as the evidence containing Talbott's DNA was most directly associated with her body, not Jay's in Snohomish County. Snohomish said fine, but a prosecutor from each county would team up to try the case as equals. And Scharf got what he wanted: he would make the arrest and take the lead in the interrogation, as it was his investigation that produced the breakthrough, while Skagit hadn't played an active role in the case in years. Skagit would send a detective to be Scharf's partner for the day.

The five o'clock hour arrived—closing time at Nelson's front office. Then came an unpleasant surprise: the trucking company rolled its front gate shut, even though a number of trucks had yet to return for the night. This was something Scharf and his team hadn't foreseen or accounted for in their plans. The gate had always been open every other time they were there.

The existing strategy called for everyone to wait until Talbott left his truck on foot and went to the work trailer or his parked car, neither of which had a view of the front gate. Scharf and his partner would then approach Talbott alone, at least so far as the suspect could see, so he wouldn't realize he was the subject of a major investigation, and therefore might be more amenable to talking. Meanwhile the backup team would creep in and hide behind trucks or cars until Scharf raised his left arm, signaling them to close in. If Talbott agreed to talk voluntarily, the team could back off. With the gate shut, everything got far more complicated.

When the next big rig approached, the outlier on the team said, once again, that it was not Talbott at the wheel. But Scharf watched closely how things worked with the gate closed. When the truck arrived at the motorized gate, it opened to let the truck pass. Scharf could not tell if it was activated from the office or if the driver had a remote. Either way, the gate remained open long enough for the truck to get far enough into the compound for the gate to no longer be visible in the driver's rearview mirror, because the Nelson property sloped downward from the gate. Even better, Scharf observed, there was enough time for him to slip through the gate, and other members of the team as well, before it rolled shut. They'd have to move fast, but the plan could still work. He updated everyone over the radio, then resumed waiting.

Shortly before six o'clock, another truck was spotted. Then came the words Scharf had been waiting for: "It's him. It's Talbott."

He watched Talbott drive up, open the gate, and pass through. Once he had rumbled down the slope, Scharf and the rest of his team moved into position. Scharf got out of his car and hid behind a parked truck. He couldn't see Talbott from there, but one of the lookouts saw the suspect leave his truck. The plan called for allowing him to get at least forty feet from his truck but still over a hundred feet from his parked car—in no-man's-land, with nowhere to hide or flee.

Scharf tensed. Then the lookout said Talbott was where they wanted him, and the detective emerged from hiding, striding toward Talbott. He carried a vinyl folder with the sheriff's department insignia on it where Talbott would see it.

"Are you William Talbott?" Scharf called out.

"Yes."

Scharf approached him with his hand outstretched, and Talbott accepted the handshake as the detective said, "I'm Jim Scharf from the sheriff's department. I'd like to talk to you."

He deliberately did not say which sheriff's department he was from. He wanted Talbott to assume he was from the department that had jurisdiction over Seattle and environs—the King County Sheriff—which would not be involved in the investigation of Jay's and Tanya's murders.

Then Scharf began his practiced spiel. His department had developed a composite of a "person of interest" in an investigation. He gave no details about the crime but said the composite had been made public and 167 tips had come in from citizens who thought they knew someone who looked like the police artist's drawing. He told Talbott he just happened to be on that list assembled from the tip line and that the department was in the process of contacting everyone on the list to rule them out. "Could you take a few minutes to talk to us about it so we can rule you out?"

None of this was true. The deception was meant to lull. But Talbott would not be so easily persuaded.

"I worked a long day and I'm really tired," the bearded trucker said with a shake of his bald head. "I just want to go home. Can you come back tomorrow?"

Scharf eyed Talbott's imposing physique.

"Well, we've come a long way to take care of this, Mr. Talbott," he said, making his voice sound a bit plaintive. "I think we could take care of this in ten minutes and then be on our way and this wouldn't be hanging over your head."

He thought about pulling out the composites in that vinyl folder, showing them to Talbott and saying, "These don't really look like you." That was in his script. Instead, he suggested they could just sit in his car and do a quick interview. Just a chat. It's not like he'd be under arrest.

The answer was still no. The truck driver was too tired and just wanted to go home and call it a night. "Maybe you can come back tomorrow," Talbott repeated. "Or the next day."

During this entire conversation, Scharf had been trying to look Talbott in the eyes, both to better get a read on him and to verify their blue color listed on his driver's license. But Talbott had kept his head tilted downward the entire time and would not meet the detective's eyes.

Feeling his opportunity slipping away, Scharf tried to engage him by asking for his identification.

"I told you who I am," Talbott replied coolly.

"I know, I just want to verify it," Scharf said, also without heat. "Please let me see your identification."

Now Talbott sounded stern. "I *told* you who I am." He looked away from Scharf.

The game was up, the detective decided. All the questions he had scripted went out the window, beginning with a long series of easy and nonthreatening ones such as: *Where did you grow up? Are you married? Who was your best friend in high school?* Then more pointed ones: *Have you ever owned a gun? Have you ever spent Canadian traveler's checks? What did you use zip ties for? Have you ever beaten anyone with a rock or choked them?* And then, finally, his closer, scripted and ready to go at just the right moment:

> *Yes, you did, you killed Jay and left him under the High Bridge right down the street from your parents' house. You left Tanya on the side of Parson Creek Road.*
>
> *You were having a lot of turmoil in your life then, weren't you? You didn't have a job. Probably had no family support. It's okay to say it now and get it off your chest. You've been a good guy since then. You will feel so much better getting this weight off your shoulders. It's been wearing on you your whole life. It will be such a relief for you to get it out. We've had people tell us much worse things than this, we won't judge you, you're a nice person.*

Scharf sighed. He had envisioned that moment. Now it almost certainly would never come. The detective raised his left arm to give the signal for the backup officers to move in and said, "You're under arrest. Turn around and put your hands behind your back."

But Talbott didn't comply. He didn't move at all, standing rigid and tense, glaring. He just said, "What for?"

Scharf looked at him and said, "First degree murder."

Now Talbott did meet his eyes. He didn't blink, react, seem surprised—or obey. Scharf braced himself for whatever came next. Flight? Fight?

Only after the backup came jogging up, a half dozen cops and more standing by, did Talbott relent and allow himself to be handcuffed. Scharf had to borrow a second pair of cuffs and link them to his own before he could successfully restrain Talbott's bulky arms.

Scharf read Talbott his Miranda rights, the ritual that anyone who has ever watched a police drama on television knows by heart, that anything he said could be used against him. At the same time, a deputy emptied Talbott's pockets. The contents were put in a paper bag. Then Scharf escorted him to a patrol car to be taken to the sheriff's office, with Talbott in back, Scharf in the passenger seat, and a deputy driving. Scharf asked Talbott again if he wanted to talk when they got to the sheriff's office. Talbott said, "Not after what you just said to me. It sounds like I need to talk to a lawyer."

And that was the last conversation of any substance he and Scharf would ever have.

As they drove, Scharf placed two phone calls to Canada: one to John Van Cuylenborg, Tanya's brother, the other to Gary Baanstra, Jay Cook's brother-in-law.

"Guess who I've got in the car with me," Scharf told each one in turn. He didn't care that Talbott could hear every word. Scharf had forewarned them that he might be making an arrest in the case. "We've got him."

The stunned family members weren't sure how to react. Relief? Anger? Grief? Joy? Scharf thought he heard all that and more in their voices.

He said there would be a press conference the next day. Both men said they'd be there, and Jay's mom and sister would come, too.

The press conference would be Scharf's chance to make a plea for friends, coworkers, and former neighbors to come forward and fill in the blanks he had hoped Talbott himself would address. He thought it might also bring some measure of peace or relief to the families, and he was glad to hear they would be present.

At the sheriff's office, Scharf brought his prisoner to one of the tiny gray windowless interview rooms. As Talbott walked to the door, he glanced at the sign on the office across the corridor: "Polygraph Room." It was 7:11 p.m.

A deputy brought leg cuffs and he and Scharf tried to put them on Talbott, but they were too small to fit around his ankles. Talbott sat in his straight-backed plastic chair, hands still restrained with the double handcuffs, and looked down at the floor the whole time, unmoving and silent. He had not said a word since getting in the patrol car after his arrest.

"Are you comfortable?" Scharf asked.

Talbott didn't look up but he did speak. "No."

"Do your wrists hurt?"

"I'll deal with it."

Scharf nodded, walked out, and locked the door.

It was still possible Talbott might decide to start talking, Scharf knew. He had been read his rights, and Scharf wouldn't initiate anything. He'd certainly listen, though. But he doubted that would happen.

A little after eight o'clock, Scharf came back to check on Talbott and explained the booking process. They would swab his cheeks for DNA, then he would be taken to the jail for booking.

Talbott, still not looking up, spoke. "I want an attorney."

"You can call one after you're booked. Okay?"

No response. Scharf left him alone again.

A half hour later, Scharf came back with Sergeant Sheahan-Lee and another detective from Skagit County. Scharf told Talbott they had a search warrant to take oral swabs of him. He held a copy of the warrant.

"Why are you doing this?" Talbott asked.

After so much silence, the question almost startled the detective. "We have a court order to get a confirmation sample of your DNA."

"Where did that come from?"

Scharf tried to catch his eyes, but Talbott still wouldn't look up.

"A superior court judge signed an order allowing us to get oral swabs of your DNA. If you don't cooperate, we can do that without your consent."

Talbott said he wanted to wait on that until he had a lawyer. Scharf read the search warrant aloud, reemphasizing that the detectives didn't need Talbott's permission to proceed, though Scharf knew swabbing a huge and uncooperative suspect would be no fun. Not in the least.

"I will not cooperate until I have a lawyer," Talbott said.

Sheahan-Lee asked if a public defender would do, and Talbott agreed. Scharf got one on the phone and she spoke to Talbott, then told Scharf she'd be there in twenty minutes.

While they waited, Scharf asked Talbott if he needed to use the restroom, if he wanted a drink, or if he was hungry. The detective got three one-word answers in return: No. No. And no.

Talbott's private meeting with the public defender lasted all of seven minutes before she emerged from the interview room and said he was ready to cooperate. Scharff swabbed Talbott while the others watched, scrubbing skin cells off his soft inner cheeks, then put the swabs into their glass tubes, bagged the samples, and let the attorney snap pictures of them with her phone. Then he escorted Talbott through the tunnel that led to the county jail.

Talbott did not say another word.

Scharf would have liked to think Talbott's refusal to talk, deny, or explain himself showed a guilty state of mind. But he knew better. Guilty crooks talked all the time. In fact, Scharf depended on those loose lips to help strengthen his cases. Given the DNA evidence, there was nothing Talbott could say that would dissuade Scharf from charging him, except maybe this: I have a twin brother and he's the guy you're looking for. Anything short of that—an excuse, an evasion, a lie, or even the smallest point of contradiction—would become a weapon for Scharf and prosecutors. On the other hand, declining to talk with a detective and asking to speak only with an attorney could never be offered as evidence of guilt, by law and constitutional right.

Scharf, who prided himself on his ability to get his arrestees to talk, couldn't help but question himself. What could he have done better? What did he miss? What explanation or defense would Talbott have offered if he had decided to talk?

They were good questions, for Talbott would indeed offer a defense in time. And he would adamantly insist he was innocent of Tanya's and Jay's murders.

22

Oh, That's Just Bill

May 17, 2018, 6:15 p.m.
Woodinville, Washington

"What's he done now?" the woman at the door blurted.

Detective Tedd Betts tried not to look startled. He'd done nothing more than introduce himself and ask for William Talbott's sister Inga. The woman answering his knock hadn't asked what happened, or if there'd been an accident, or if Bill was okay. She had gone right to "What's he done now?" To any detective, this would be a promising start.

Only fifteen minutes had passed since the cuffs snapped around Talbott's wrists, and, per Jim Scharf's plan, Betts was there to interview Inga before word got out. But the woman at the door said Talbott's sister wasn't home. She had gone to visit her husband at the hospital, where he was being treated for cancer. The woman told Betts she was Inga's mother-in-law, Ruth.

Something about her assertive posture and tone told the detective his best bet would be to answer Ruth's question honestly. "William Talbott has just been arrested for homicide."

The mother-in-law didn't hesitate. "I'm not surprised in the least," she said. Even more promising, the detective thought. Scharf would be dancing a jig when he heard this.

"Can we talk for a little bit?" Betts asked.

So the detective in his plain clothes, badge on his hip, and Snohomish County Sheriff's ID around his neck, sat on the porch with the mother-in-law and two other detectives.

Billy, as she called him, was a secretive man, always had been. He cut ties with his family long ago. He drifted from dwelling to dwelling, job to job. He had lived for a time in Inga's basement in a makeshift bachelor's pad. It hadn't ended well. Even when he was still talking to his sisters, he kept things hidden.

Talbott had been bullied in school, Ruth said, "probably because he was so annoying." Maybe that was why he dropped out in tenth grade and never returned.

On the other hand, she said, she had never observed him being violent, or with a gun. But there was a story Inga had told more than once from when they were kids and she found herself alone in the house with Billy: "He pinned her down and felt her up."

Ask Inga, Ruth said. She'll tell you all about it.

Betts suggested she call her daughter-in-law and ask her to come home. Just say the police wanted to talk to her about Billy, the detective said. But don't tell her why.

A half hour later, Inga pulled into the driveway. When Betts explained he wanted to talk to her because her brother had been arrested for murder, she gasped. Then she fished out of her purse a note she had jotted down at the hospital. On it was a list of the things she thought her brother might have done to bring the police to her doorstep. She handed the paper to the detective, then opened the door and said, "Let's go inside."

He looked down and read what she had written: "Assault, Fraud,

Arson, Rape, Murder, Robbery." Next to that list she had scrawled the time and date, then, as an afterthought, she had added another option: "Dead."

A TANTALIZING, CONFUSED portrait of the man behind the DNA began to emerge for Jim Scharf, a Swiss cheese biography as told by his two sisters and his father, pocked with holes no one could fill, thanks to Talbott's practiced silence and many years of withdrawal. About fifteen years ago, maybe a bit more, he had cut off ties with his family. Not right after the murders, as Scharf had expected, but more recently. His sisters hadn't talked to him since the early 2000s. He had lived next door to his eldest sister, Angel, and he'd walk right by her as if she wasn't there. None of his family could say why. The only relatives he'd speak to were his nephew and niece—Angel's kids. But the rest of the family had plenty to say about him in the years that preceded the break.

The stories were at once disturbing and heartbreaking, pitiable and grotesque, and most of all contradictory. They spoke of a child and man at times consumed by anger yet desperately seeking approval, a bright kid who failed at school, possessed of a wild temper but desperately eager to please—when he wasn't being maniacally opinionated. He was epically unlucky at romance, awkwardly eager and painfully ingratiating with potential girlfriends, a chunky Uriah Heep—when sober. After a few drinks too many the façade would slip and he'd become offensive and mean with women. Sometimes two beers were all it took. At times he worked tirelessly at household chores or auto-repair projects as favors for friends, then he would display a slacker work ethic for his paying job, even when his livelihood and a roof over his head were on the line. And when his anger would spill over into threats and violence at home, his mother, Patricia, could be counted on to minimize and dismiss the warning signs of worse to come, telling his sisters and her husband, "Oh, that's just Bill." He's just blowing off steam. He's just going through a stage. He's just depressed again. That's just Bill. Those words became the household mantra.

The warning signs started in the sixties, when the family lived on what was then a semirural road in the town of Woodinville, in a home with acreage and horses, goats, dogs, cats, and a large tom turkey. At the time, Woodinville was just another forest town outside of Seattle. What would become the town's most famous corporate citizen and tourist draw, the sprawling Chateau Ste. Michelle vineyards and winery, was then a fledgling label with a couple of wines no one had heard of.

One day, when Billy was six or seven, he spied one of the family cats, Nicky, walking on the edge of the family well, the source of household water at the time. Billy pushed the pet off the edge and into the well. His father had to rescue the screaming cat and then have the well sterilized. He was furious with his son.

On another occasion, young Billy reportedly threw one of the family cats into a partially frozen lake.

His grade school years were difficult. "He was picked on at school mostly because he didn't know when to quit," Inga said. "He'd be an irritant to a lot of kids."

"Bill was intelligent with no social skills," sister Malena offered. "So he'd get bullied at school."

Several friends, however, mentioned something the family didn't bring up: that at some point, Billy developed a stutter. They said it was very noticeable in his teen years, and they thought that was part of the reason he was bullied.

At home, his sisters said, he was the bully as he grew older and stronger and his "anger issues" became more pronounced. "He was angry and unhappy a lot, and the family would just tiptoe around him," Malena told detectives.

On September 24, 1974, disaster rocked and forever changed the Talbott family, sapping whatever capacity they had to help Billy—or anyone else—deal with home, school, or personal crises. Bill Talbott Sr. had taken his brother for a spin on his motorcycle that day, cruising down the Woodinville-Duvall Road. The family house sat on that road, the very same thoroughfare that led to the High Bridge turnoff, a winding, leafy joy on a bike. Bill Sr., then thirty-four years old, was an accomplished motorcycle enthusiast. He knew that road well, as only a local could, had made countless pleasure rides, trips to the store, and his daily commute to work. But when a white Toyota heading in the opposite direction crossed the center line in front of him, no skill or familiar terrain could save the day.

That they both survived the terrible collision seemed a miracle, but the injuries to both Talbott brothers were devastating. Spinal injuries left his brother paralyzed for life from the waist down, while Bill Sr. suffered multiple fractures and dislocations. His arms were so shattered that the first surgeon to look at him recommended amputating both. Bill Sr. said no, but the alternative wasn't great, either: four years in and out of the hospital, fourteen surgeries, months in an "airplane cast" with one arm suspended like a broken bird's wing, and, of course, life-changing, never-ending pain. In the end, he was left with one arm mostly paralyzed and the other limited by a fused wrist.

His ability to work, and to do just about anything else, was severely affected, and the kids had to pick up the slack. Angel was thirteen at the time of the crash. Billy was eleven. Inga had just turned ten. And Malena, the baby, was seven. As she recalled, "We all had to grow up in a hurry. It wasn't a typical childhood by any means. We had to take on bigger roles than we probably were prepared for. Me helping my mom feed my dad. Bill had to take on things that Dad had done. We just all had to kind of step up and take care of each other."

This would be a potential breeding ground for anger, resentment, jealousy, and dysfunction in any family. It did not seem to bring out the best in Billy. Instead of directing his anger at helpless animals, he began turning it on the rest of the family—everyone but Angel. He didn't dare pull anything when she was around. Not only was she the oldest, she was the toughest. In junior high she was a soccer goalie—on the boys' team. Billy watched his step when she was around, the other sisters recall, but when he was alone with one of his younger sisters, things happened.

About a year after the crash, while their father struggled with recovery and more surgeries, Billy barged into Inga's room. He was twelve, she was eleven, and he pushed her down on the bed, got on top of her, and started touching her. She was just beginning puberty, self-conscious and confused about the changes in her body.

"I didn't even know what it was because nobody's ever done that to me before," she told the detective. She kept trying to push her brother off, shouting at him, but he was bigger and stronger, already starting to take on the burly form he'd grow into as an adult.

"Why are you being so weird? Stop being such a weirdo!" she called out. Repeated use of the word "weirdo" finally seemed to get through to him, and he relented and left the room. She felt angry, confused, ashamed, and violated. But when she complained to their mom, Patricia told her not to make a big deal out of it. Bill was having such a miserable life, she said. Girls didn't like him. He just was curious, just wanted to feel what a girl's parts were. She realized her mom was defending him, not defending her, the one who was attacked, and she protested.

"Oh, that's just Bill," her mom said. "It's just what boys do when they don't have anybody. He just got on you to feel what a girl's boobies were like."

A couple years later, Billy had a shoving match with his hobbled father, arguing over some chore his dad wanted done that Billy didn't want to do. The stitches from his father's latest surgery split and soon his shirt was covered with blood and everyone was shouting at Bill, who fled. Later, sputtering with anger, Bill told his dad to just wait until he turned sixteen. He was going to get his license and the first thing he'd do was run his father over.

"Then you won't get your driver's license until you can get it yourself," his father shot back. True to his word, the elder Talbott refused to take him to get his driver's test and license. And Talbott never did get his license until he turned eighteen and could go to the DMV himself.

One of his most annoying habits as a teenager was his love of dismantling home electronics and appliances, trying to figure out how they worked. The problem was, he either couldn't or wouldn't put them back together. So he'd leave them in pieces. It drove his father crazy. They'd go to turn on the radio or turn on a lamp, and it wouldn't work. Billy had gutted them.

Complaining to their mother was pointless, though the girls kept trying. Inevitably, their mom would start to say, "Oh, that's . . ."

"Just Bill," the girls would finish. "Thanks a lot, Mom." They resented being held to a higher standard than their brother.

AND YET, BILLY had an incisive, intelligent mind. This was yet another infuriating thing about Billy Talbott. The general consensus in the family was that he could have been the star student of the four kids. He picked up things with little difficulty. He would pass tests in most subjects in school easily, but he would still get failing grades in class after class because he refused to do any of the work. He finally dropped out with an abysmal report card in tenth grade. His teachers urged him to get his GED and go on to college, but his attitude, according to friends and family, was that if he was that smart, he didn't need any more schooling. He would just read voraciously and learn that way.

After he quit school, his father expected him to earn his keep at the house, but when he shirked one too many tasks for his dad to bear, he threw him out. At fifteen, Bill was living behind the house in an old trailer.

One day he snuck into the house when no one was home and listened to music in Inga's room. He preferred to blast the stereo at peak volume, loud enough at times to cause neighbors down the road to complain—in a rural area where the houses were spaced far apart. He wasn't even supposed to be in the house, and when Inga came home and needed to change to go out, she locked her door and turned down the music. He started banging on her door and shouted, "Turn it back up!" Enraged when she ignored him, he broke the door open. He shoved her out of the way so hard she flew backward and fell, breaking her tailbone. Bill ran off and hid in the woods, leaving her on the floor crying in pain and anger. Her parents had to rush Inga to the hospital. They did not involve the police.

The parents took all the kids to counseling after that, but little seemed to change at home. "He started getting angry all the time," Malena said. "Like life owed him something, that he lost his childhood."

However, another side of Talbott emerged in his teenaged years when a new family moved a few houses up the road and the new neighbors took him under their wing. Marilyn Seat was the kindly aunt Bill felt he never had, her house always full of kids and extended family hanging out in the big field next door, riding and working on motorcycles, barbecuing. There was always something happening there.

Seat's younger sister, Joyce, told detectives how Billy introduced himself in the spring or summer of 1980, when he was seventeen. He pedaled up on a ten-speed bike, then squeezed the front handbrake hard and flew over the handlebars. Joyce's kids and their friends laughed, along with Marilyn's nephew, Mike Seat. Talbott popped right back up and grinned, unhurt. And then he hopped back on his ten-speed and did it again.

"Why'd you do that?" Joyce asked.

"Well, you guys laughed. I'm Bill Talbott. How ya doin'?"

And now Joyce joined in the laughter and invited him to come on over and meet everyone. Pretty soon, Talbott and Mike Seat had become friends. Seat was nearly three years older, but he felt bad for Talbott and found him to be good company once you got to know him—and once you learned the best way of dealing with his habit of arguing about everything was to not argue back. They would hike or ride together, and when they were a few years older, they would work together and room together.

Once he had a license and car, Talbott started showing up at Joyce's house in the city of Snohomish, where he would help her husband, another Mike, with chores, or working on his motorcycle, or rebuilding an old pickup engine.

"He was like a lost pup," Joyce told detectives. "He'd just show up."

She felt sorry for him and got the impression his family members were the ones who bullied him. But she also found him wearying to be around, argumentative and opinionated in a way that drove off the girls he would have liked to ask out. Instead of modifying his behavior, he came up with a scam. He'd figure out what a girl he liked from afar enjoyed doing, claim that was his passion, too, and frantically study up on it. The first time Joyce noticed him use this tactic involved a young woman who was into white-water rafting. Talbott persuaded Joyce's husband and son to join him on a rafting trip, then invited the girl along. It was a disaster. Talbott alternated between fawning attention and belligerent opinions about proper rafting technique. After the trip, the girl wanted nothing to do with him. Joyce saw this pattern repeat several times. He'd try to woo a crush by misrepresenting himself or his hobbies, followed by the inevitable crash and burn. He'd mope and lick his wounds for a month or two, then do it all over again.

Joyce tried to explain why she believed he continually failed in forging a relationship. She spoke hesitantly, not wanting to sound hardhearted, but in the end, she had to say it: Talbott was always his own worst enemy.

"It isn't very nice," she said, "but he was like a double nerd. He didn't think he was, but in reality, he was. He was opinionated. . . . You couldn't talk to him, you know? He just wasn't the lovable kind, let's put it that way. He just wasn't. Wanted approval all the time, but then wouldn't back down. We argued because I wouldn't listen to his BS. . . . He'd try to do jokes, and he'd blush and stuff. I felt sorry for him . . . that he didn't have somebody."

A little more than a year after the Seats moved to the neighborhood, Talbott had fixed up an old motorcycle and was out riding—illegally, given that he was seventeen and still had no license. A neighbor backed out of a driveway without looking and, in a horrible echo of his father's crash, the younger Talbott slammed into the car that suddenly materialized in front of him.

The crash was not nearly the accordion impact Talbott Sr. had endured—Billy Talbott's was a ninety-degree collision, not a head-on, and the closing speed was less devastating. Even so, the damage was serious: a broken knee and a fractured pelvis, among other injuries. Though he would fully recover, Talbott was in the hospital in Everett for nearly a month, turning eighteen in a cast and a hospital bed. His father let him move back into the house to complete his recovery, and for the first time in a long time, there was a stretch of family life without discord. It was more a ceasefire than a peace treaty, but the battling Talbotts took what they could get.

Then the younger Talbott accepted an insurance settlement. The driver who backed into Talbott's path admitted it was all his fault for not looking carefully enough for oncoming traffic. The settlement covered Talbott's medical expenses and put twenty thousand dollars in his pocket. That would be sixty thousand in 2021 dollars, and it was, in any era's dollars, the most money William Earl Talbott II had ever had in his life.

He immediately moved out of the family home and rented an apartment in Everett, then put a down payment on a flashy silver Mercury Cougar. In short order he had a girlfriend, too, for whom he had nurtured a hopeless crush for years. Suddenly, they were inseparable. They spent the settlement profligately, and when the money ran out, the smitten Bill Talbott woke one morning to find his girlfriend gone. Broke and depressed, his car repossessed, he moved back in with his parents.

He became an avid hitchhiker when he didn't have access to a car anymore, traveling far and wide. At least one friend, Michael Seat, said he loved to ride Seattle's many ferries, too, and that he particularly enjoyed hanging out in Bremerton.

In 1984, when Talbott was twenty-one and Malena was seventeen, he kicked her so hard with his heavy boots she called the police. Talbott was arrested for assault. This was the arrest Scharf came across when he first started backgrounding Talbott. Malena recalled that the detectives at the time photographed her legs to show the severity of the injuries, but the case was ultimately dismissed and Talbott paid a $150 fine. He eluded a criminal record.

The spark that sent her brother into a violent rage: she turned down the volume on the television.

We always had to forgive Bill, Inga told detectives. "But he never forgave anyone for anything."

His father ultimately kicked him out again after an argument over chores Bill was supposed to do. He had gone fishing instead. After that, he couch-surfed at friends' homes and lived in a camper on a friend's property. The friend also gave him a job with his construction company.

Bill moved from job to job, then tried self-employment, hiring himself out as a construction worker or handyman. He bought a box truck and fixed it up, then converted it to run on propane, a rolling advertisement for a propane tank delivery service he started in a region where propane was a vital fuel. Even so, it was a spottily profitable business at best, and Talbott's habit of impulse shopping often left him broke when the monthly bills came due. In lean times he lived in the truck, bundled under stacks of blue packing blankets in cold weather— blue blankets similar to the one found shrouding Jay Cook's body under High Bridge. His father remembers him having a collection of camera equipment in the back of the truck; he loved taking pictures. His friend Joyce said he was the wedding photographer for her daughter's wedding in the early 1990s, and he spent five hours there trying to get good images. He used a camera with removable lenses of different sizes, she recalled, and it looked like Tanya Van Cuylenborg's, but it could have been any single-lens reflex 35 millimeter camera of the era. She didn't remember the brand.

In 1986, Billy showed up at his sister Inga's house, fallen on hard times. She agreed to let him live in her basement in a semi-separate bachelor's pad. She recalled that he moved out when he got a good, steady job truck driving for an aerospace parts manufacturer, and he moved in with Mike Seat and another coworker. He returned after losing the job and getting kicked out by his roommates because he couldn't pay his share of the $475 monthly rent. Inga's recollection put that around the time of Tanya's and Jay's murders in November 1987.

Through all this tumult, from childhood to adolescence to young adulthood, there was one steady, loving, supportive presence in Bill

Talbott's life: his maternal grandfather, Edwon Peters. Peters was the most accomplished person Talbott knew. He had been a successful insurance agent in Seattle who invested his savings wisely, buying up homes and real estate all over the area when the economy was in decline in the early seventies. He lived comfortably off the rental income after he retired, his properties accruing value over time. He even bought a little Cessna plane and learned to pilot it, which impressed his grandson to no end.

For many years, Talbott's granddad took him camping and hiking, teaching him about the outdoors and how to enjoy the wilderness, while also gently schooling him in the virtues of discipline, planning, respect for nature, and courtesy to fellow nature lovers. The trips gave Billy an escape from the tensions and battles at home, while also giving his parents and sisters a much-needed respite. He'd often stay with his grandfather when the weather was too bad for a wilderness trek. Instead, Talbott would lose himself in Edwon's library of books. The kid who hated schoolwork loved learning from the history and science books at his granddad's house, or the literary classics—anything, really. If his grandfather liked it, that was good enough for Billy Talbott. Sometimes they'd watch a science program together on TV, one of the Jacques Cousteau undersea exploration shows that were a staple on network television from the sixties into the eighties.

In later life, Talbott would tell his friends that he learned a lot from his grandfather, and that his house was a safe haven from a terrible home life.

The two often went on camping trips together on the Olympic Peninsula, sometimes flying there in the little Cessna instead of driving. A favorite camp destination was Hurricane Ridge in Olympic National Park, where they could go hiking for two or three days at a time. The park's main entrance was in Port Angeles—the city where Jay and Tanya arrived via ferry on their fateful trip from Canada.

Talbott was familiar with that area, with the roads that Jay and Tanya traveled, and with the ferry terminals in both Port Angeles and in Bremerton, where they bought their ticket to Seattle.

After his grandfather died at age seventy-two, Talbott helped his grandmother and parents clean out the house. He got to keep his grandfather's treasured camping gear and his old red station wagon, as Talbott had no car at that time.

Grandfather Edwon Peters died on November 24, 1986. His wake and funeral were over the Thanksgiving weekend.

Tanya Van Cuylenborg's body was found on November 24, 1987, exactly one year to the day from Grandpa Edwon's death. Jay Cook was found on Thanksgiving Day.

Bill Talbott lost his truck-driving job and his home shortly before that one-year anniversary of his grandfather's death. An engine fire had killed his grandfather's old red station wagon, a treasured link to the man he loved. At loose ends, had Talbott decided to mark the one-year anniversary of Granddad's passing with a return to their old stomping grounds on the Olympic Peninsula? Without a car, had Talbott taken the ferry to Bremerton? Had he hitchhiked across the peninsula? Had he gone to a beloved landscape looking for solace—or a way to express his anger over his loss and grief, and for a life that seemed to have hit rock bottom? Is that how he crossed paths with Tanya and Jay? On the road or on the ferry to Seattle?

The coincidence of dates and location is so great it seems impossible to ignore. And probably impossible to prove at this late date, unless Talbott decided to talk.

For all his anger and violent behavior toward them, one member of his family after another, as well as his friends, told the police the same thing: they would never have believed he could kill in cold blood, not the way Jay and Tanya died. Even Inga, who wrote that odd list predicting murder as one of the reasons the police might be interested in her brother, said she didn't actually believe him capable of that crime.

"I mean, he'd be a bonehead, but not to the point of attacking and killing somebody," Malena said. "I could not picture that. I honestly cannot think of anything that would make me think Bill would do something like this."

Even more striking is how Talbott behaved after November 1987—after the murders of Tanya and Jay.

Scharf had expected this time frame would prove to be the beginning of Talbott's withdrawal from the world, the time when he started cutting ties with his family. But no. Just the opposite. He was not depressed. Not withdrawn. Not more violent and angry than ever. He was a changed man for the better, they said.

Suddenly, right around this time, and for the next fifteen years or so, those who knew him best saw Talbott as a more positive, hardworking, pleasant, and, for the first time anyone could remember, engaged member of the family. His work ethic became stellar. He helped around the house. He played with his sister Angel's kids. He helped his dad. He came to holiday dinners that he had previously avoided. He ended up living at his sister Inga's house for years, until 1990 or 1991. His nieces and nephews all called him their favorite uncle. He was the fun one.

"Late eighties, early nineties, I remember him coming by and saying, I'm so grateful you guys still want me around," Malena recalled. "Because, you know, I was such a brat as a kid."

He had finally grown up, it seemed, and become the brother and son his family always wanted. They didn't know why he had changed so abruptly and so dramatically, but it was unmistakable, and they welcomed him back into the fold.

Perhaps this was just coincidence. But if his DNA truly revealed him as a murderer in November 1987, then the timing of this sea change in personality and lifestyle would be beyond chilling. For it would mean brutally beating and murdering Jay Cook, then raping and executing Tanya Van Cuylenborg served as the prelude, if not the catalyst, for the best times of William Earl Talbott II's life.

For the first time in a very long while, he seemed truly happy.

IT WOULD BE almost a week before Jim Scharf could meet with the detectives on the interview teams and begin to digest the mass of information they brought in from the Talbott family. First the detectives who did the interviews had to write their reports, and Scharf had to get the recordings transcribed. Meanwhile, he had a press conference to deal with on the morning after the arrest.

The meeting room was packed with reporters and camera crews from local, national, and international media. It was different from the month earlier, when the Golden State Killer arrest had drawn global national attention primarily because the culprit had been such a prolifically violent serial rapist and killer who seemed uncatchable. At that time, the genetic genealogy angle had been just an added little bonus, a fascinating but complicated backstory, something exotic and probably a one-off. The experts were saying there'd be no rush to solve other cases that way, because the time and manpower needed was prohibitive. The Golden State Killer was depicted as a moonshot-level effort, and cops only have so many of those in them. And this seemed to make sense. But only if you didn't know just how much one CeCe Moore on her couch with an open laptop could accomplish.

The story flipped in a hurry with the press conference on Jay and Tanya. Suddenly it was not so complicated, after all: a genetic genealogist got involved for the first time with the case just a couple of days after the Golden State Killer's arrest, then solved a thirty-one-year-old mystery in *two hours*. The revolutionary nature of this accomplishment became instantly clear: this was the wave of the future, neither too exotic nor too hard nor too slow. Genetic genealogy wasn't the backstory anymore. It *was* the story.

"What an amazing world we live in today," Laura Baanstra told the gathering. Her voice shook but her eye contact with the TV camera in front of her did not waver. Behind her was a screen with a live feed from California of CeCe Moore, removing her big glasses and wiping tears as she listened to Jay's sister. "It's hard to put into words the feeling of relief, joy, and great sorrow that this arrest brings. The hole that was left in our hearts will never be filled completely, but the work that was done here by these incredible, hard-working professionals, both now and thirty years ago, has helped make the hole smaller."

"It if hadn't been for genetic genealogy," Jim Scharf added, "we wouldn't be standing here today."

Then the detective made a plea for anyone who knew Talbott, particularly in the eighties, to get in touch. Asking this showed the limits of the magic of genetic genealogy. It was an amazing suspect identifier, but it didn't just conjure a complete investigation. Scharf still had to do the detective work to build his case: amass usable evidence, hunt down witnesses first interviewed in 1987, and try to drum up new ones who might know something about Talbott. He announced that he was especially interested in anyone who saw Talbott back then with a Minolta camera, a blue blanket with a slit cut in the middle that may have been worn as a poncho, or a large copper-colored Ford van.

The media attention on the case in Washington and Canada made it especially hard for locals to miss, from morning TV to late-night news, public and commercial radio, and on the front page of most newspapers in the state. Many newscasts and articles featured color photographs of the evidence Scharf sought, particularly the van, that sweet, heartbreaking snapshot with Jay Cook and his mother standing beside the hulking Ford, Lee Cook striking the walk-like-an-Egyptian pose from the old Bangles song, number one on the charts in 1987, although many reports continued to misidentify her as Tanya.

A FEW DAYS later, halfway across the state in the town of Entiat, a former roommate of Bill Talbott's by the name of Michael Seat answered his phone and heard his aunt Marilyn on the line. "Guess who got arrested?" she asked. After a few seconds of silence, she started to provide the answer: "Your old friend Bill—"

"Billy Talbott!" Seat exclaimed before she could finish. He was not the least bit surprised. The Talbott he remembered hanging with back in the eighties was always up to something shady, scamming or thieving or reneging on debts. Nothing serious, some of it in a gray area, but none of it to his credit. And now Seat figured his old ways had finally caught up with him. He asked his aunt, "So what was he arrested for?"

"Murder."

"No way!" Seat shouted. "My god."

"I don't believe it, either," his aunt said. She had been fond of Billy Talbott since she knew him as that goofy teenager hanging out in front of her house in Woodinville. Seat's long friendship with Bill had withered ages ago, but his aunt still had a soft spot for her nephew's former roommate.

Seat tried to read about the arrest online, but his internet connection at best moved like the Seattle rush hour traffic he had fled when he moved to Entiat. Money was tight. Heart disease had forced him to retire early from his work as an auto mechanic. He played country and rock in a band called Last Chance, about the only thing his doctor would let him do. After staring at one too many web pages that wouldn't load, he drove over to his mother's house to use her broadband. Sitting at her computer, he started reading news stories about double murders and dead Canadian sweethearts and cold cases and genetic genealogy. He gathered that the murders had been big news back in the day, though Seat couldn't remember anything about the case. He didn't read much back then—life in his twenties revolved around fixing and racing cars and motorcycles, the interests that had first brought him and Talbott together. None of what he was reading made any sense to him. So far as he knew firsthand, Talbott never did anything worse than file a phony auto insurance claim or treat girls like shit when he was drunk. Yes, he knew Talbott had a temper, and Seat had heard some stories about altercations with his sisters, but Seat personally had observed Talbott express anger with words or silence only, not violence. He just couldn't see the Billy he knew as a killer. No way.

Then he saw the photo of the copper-colored van with the big shiny hubcaps on the screen. He gaped. His hands began to shake, hitting random letters on the keyboard. Dizziness almost tipped him out of his chair. He closed his eyes, yet he could still see the van, a flash of sunlight on chrome. A memory had pierced him like an arrow. He wanted to throw up.

"Michael?" his mother asked. "Are you okay?"

Seat looked up and his mother saw his wide eyes and pale face. He tried to answer, but both his voice and his shrunken, myopathic heart stuttered at the same time and nothing came out. He cleared his throat and tried again.

"Yeah, Mom, I'm okay," he croaked. "Mom, I seen that van. That van was sitting in the driveway one morning when I drove by back then. I seen it plain as day." He closed his eyes, then said, "I can still see it."

His mother asked what in god's name he was talking about. Where did he see that van?

The moment replayed in his mind. It was on a loop now, he couldn't stop it: driving that road he knew so well, winding through the autumn woods early in the morning, his old silver Camaro scattering the last lingering shreds of mist, then emerging into sunshine. And there was the van, the hubcaps glinting in the sunlight, and . . . No. He didn't want to remember any more. Not the bad part.

"I saw it parked on the Woodinville-Duvall Road, Mom. Where Billy Talbott used to live. I saw that van in his parents' driveway, right next to the blackberry bushes."

Silence. Then, finally, his mother asked what he was going to do.

"I don't know."

"Yes you do, Michael. Call the police." She reached over and scrolled down one of the articles to the end, where the police hotline number had been displayed.

He nodded. "I better go home and call."

He left, unable to voice the really bad part, not even to his mom, the sinking feeling of horror, the fear that still gripped him as he drove home. It was not just that he remembered this key piece of information from long ago. Or that his former friend might be a killer. It wasn't even the realization that people would have a hard time believing this thirty-one-year-old recovered memory—at least, if they hadn't witnessed him nearly pass out as he remembered it. No, it was the timeline he had just read in one of the articles that tormented him. That timeline showed that three days had passed between the time those kids disappeared to when the van had been abandoned in Bellingham. Three days. That's what was killing Seat. Talbott had dumped Jay's body at High Bridge first. Then he drove the van to Skagit County and left Tanya. Then it was a short drive to dump the van in Bellingham.

Didn't that mean Tanya was in the van when Seat drove by and saw it? Oh my god, did that mean she was still alive then? What if he could have done something? He had been curious about the van back then; he had made a mental note to ask Talbott about it. So why oh why hadn't he just pulled over and asked about the van then and there? He could have stopped Talbott. Maybe he could have saved Tanya. At the very least, he might have found the body and saved those poor families decades of pain and fear.

He parked his car in the driveway and walked into his house a short while later. He had to sit down, head in his hands, and try to stop the rush of nausea roiling his gut, try to convince himself there was nothing he could have done, no way he could have known. As he tried to calm the lopsided thump in his chest so he could call the police and coherently explain what he remembered, the phone rang, making him jump. Jesus, he said to himself. Get it together.

He picked up and said hello and heard a man's voice say he was a detective from the Snohomish County Sheriff's Office, calling for Michael Seat. The detective's name was Jim Scharf.

23

Heart of Gold, Van of Copper

Bill Talbott wanted out. Desperately. He complained to his friend over the jail pay phone every few days. He complained about the jail food. He complained about the rules against pinning up photos of women in low-cut tops. He complained that he couldn't use his sleep apnea machine because the cord wouldn't reach.

"They treat you like shit in here. Supposed to be innocent until proven guilty, but there are no rights in here. Some of these guards even make you wait for breaks while they sit there and bullshit."

"I'm just surprised they are continuing to go through with this whole fucking menagerie," the friend said—the friend who had been taking care of Talbott's house and bills, the friend who had been rounding up letters of support for Bill, the friend who said he'd take a bullet for Bill. He was also the friend Bill promised would inherit everything he owned if something happened to him in jail. "It doesn't make sense why they would continue to try when they don't have anything."

"That's why I have to make bail," Talbott said.

His lawyers were making a big push for release on bail, asking that he be able to leave the high-security jail cell surrounded by hardened felons who gave him the mocking nickname "Cold Case."

Talbott had been targeted by other inmates, suffering at least one unprovoked attack. Because guards noted that he did not fight back, the jail did not enforce the usual policy of putting everyone involved in a physical altercation into solitary confinement regardless of who started it. His lawyers cited this good behavior as evidence he'd be a good candidate for release, and they argued that his deteriorating physical condition added urgency to their request. Talbott's health in jail was poor, and he had begun losing weight at a rapid pace.

Defense lawyer Rachel Forde told the court, "Mr. Talbott has no reputation or history of violence. This accusation alleges behavior completely divorced from the responsible manner in which Mr. Talbott has lived his life."

When Scharf read Forde's statement, he grumbled to one of the prosecutors on the case, Justin Harleman, that Talbott's family told a different story. Scharf wanted to be very clear that under no circumstances could Talbott be released. Harleman told him not to worry, that there was no way that would happen. Not on this case. But Scharf had seen stranger things with some of his other cases.

Letters to the court supporting Talbott and advocating for his release flowed in. They described a man starkly different from the portrait offered by Talbott's family in those initial interviews with the police, though they were in keeping with his sisters' account of a man changed for the better in the years following the murders.

Typical was Ardell Klaus's handwritten note describing him as a family friend for more than a decade, spending every major holiday with her and her family. "Bill is a very respectful man and has a heart of gold. Bill is very hardworking and a people pleaser. Bill spends most weekends with us at our property helping out with any chores that need to be done without complaint. . . . Please don't see him as an evil person because he absolutely is not, he has a special place in my heart as well as many others."

Fellow trucker Richard Klaus called Talbott his best friend for nineteen years and a "loyal, caring, loving and unselfish person" who was "the first person to 'gather the troops' to help someone in need." Larry Williams, a friend for twenty years who worked construction with Talbott, couldn't imagine him ever striking or even yelling at anyone, calling him "kind-hearted and gentle with my children, grandchildren and even my animals"—a man he "trusted without reservation." Longtime camping buddy Jason Faaborg swore he never saw him raise his voice, get angry, or let a disagreement escalate to an argument. "I still do not believe this man could have done this terrible crime or even participated in any way. People [capable of that] just do not have the peace and temperament that Bill exudes."

The letters uniformly described a man who had become like family; trusted with watching homes during vacations, beloved by children, and a reliable source of support and comfort in times of need. It is not unusual for a few friends or family members to write letters to the court pleading for release or mercy in sentencing, but this number—and the almost saintly characterizations of Talbott— was extraordinary.

None of the letter writers seemed to have known Talbott at the time of Jay's and Tanya's murders. Almost all had met him in the late 1990s or early 2000s—what his sisters called his best years—and apparently he had stayed close to them after he had cut himself off from his family.

His father and sisters did not write to the court, although the teenaged daughter of Talbott's late sister, Angel, did submit a letter fondly recalling bicycle and motorcycle rides with her uncle and assuring the court the Bill Talbott she knew could not have committed such terrible crimes.

The prosecution submitted no testimonials. Deputy Prosecuting Attorney Harleman instead argued that Talbott was good at hiding his true nature, and that his behavior with friends did not predict his behavior toward other people.

"The only way a person could commit crimes such as these is if they indeed have a natural inclination for violence. That inclination may not come out at holiday parties or while camping with friends, and it may only rear its ugly head once every thirty years, but that is often enough to put the court on notice."

The judge agreed. The charges were too serious, the flight risk too great. Talbott would stay in jail until the jury had its say.

It might be easy to dismiss these testimonials as part of a concerted campaign engineered by the defense lawyers or one determined friend with his eye on a payoff. But there were others who were not part of this campaign, who had little to gain and much to lose from an association with Bill Talbott, who nevertheless had nothing but good things to say about the man. Scharf and his partner had found another person who knew Talbott well and who spoke of him in similarly glowing terms, and this witness had particular credibility for the detectives. He was one of their own, a former Snohomish County Sheriff's deputy.

Mike was a navy pilot stationed in Bremerton, Washington, when he first met Talbott in 1991. Talbott moved into a rental attached to the naval officer's home, where the trucker lived as both a tenant and close friend of the family. After leaving the service, Mike became a cop, a town council member, and then eventually moved to the East Coast for a job in federal law enforcement. Talbott lived with the man's family for about ten years and even did drywalling with Michael Seat on a home addition and garage on the property.

"Bill is a talented tradesman who took great pride in his knowledge of work," Mike wrote in an email to a Snohomish County Sheriff's detective. "His detailed attention to rules as a professional truck driver is a dichotomy having committed this devastating act; he was very critical of people breaking rules as a professional truck driver."

Mike and his wife were stunned by the news of Talbott's arrest. And they were devastated by the realization that someone who could commit such a crime had lived with them and had been part of their extended family for so many years, without them ever having a clue.

"I was a very good friend from his perspective and I believe that our family served as an emotional replacement for his real family, creating what he said . . . was the most stable time of his life," he wrote the sheriff's investigators. He told them Talbott was very good with kids and "was generally friendly, personable and jovial, but would become depressed at times."

Even after the arrest and learning about the evidence against him, Mike suggested the crimes might not have been planned in advance but had been a more spontaneous series of events that spiraled out of control.

"Bill is a linear thinker in general; he tends to make short-sighted decisions. This incident may have started with one thing and then a series of linear decisions which kept getting worse and worse."

This was too much for Scharf. He was convinced that Tanya and Jay's killer had been on a mission when he crossed paths with his victims and knew exactly what he would do, equipped with the kit he needed to do it. But the image painted by his friends, supporters, and former landlord meant the story of Bill Talbott was not simple, not all darkness. There clearly was another side to Talbott's nature, and it appeared to be in stark contrast to the person accused of a brutal rape and two murders.

That didn't mean Scharf felt the slightest sympathy for the man.

But it did worry the detective, for it suggested that Bill Talbott might not be so easy to convict.

JIM SCHARF HAD only a general idea of what he would learn when he and his new partner, Detective Joe Dunn, drove the three hours east to Michael Seat's home in Entiat.

His phone conversation with the man had been brief. Seat had seemed flustered and exclaimed, "I was just about to call you!"

They had discussed a few details—enough to establish that Seat needed to be interviewed at length, in person, and on tape—and carefully walked through his long association with William Earl Talbott II.

Michael Harvey Seat was fifty-nine when Scharf met him, but he looked older, with a monkish fringe ringing his bald head. Seat had dark, puffy pouches under his eyes and he sometimes sounded short of breath. He looked ill to Scharf, which Seat confirmed, telling him about his cardiac myopathy, his pacemaker, and the fact that if his heart got any worse, his doctor would put him on the transplant waiting list. Talbott's old neighbor, Joyce, had provided a snapshot of Talbott and Seat together from the early 1980s, showing Seat with a big head of bushy dark hair and a grinning, rakish look about him that heart disease had stripped away.

Scharf could see right away that Seat was not the sort of person who clammed up when nervous. Quite the opposite. He happily launched into reminiscing about fleeing Entiat on his motorcycle the day Mount St. Helens exploded in May 1980 and staying awhile with his aunt in Woodinville. He recalled meeting the funny kid from down the road who careered to a halt on his bicycle and did flips over the handlebars.

He had visited the Talbott family home on the Woodinville-Duvall Road and met Bill's parents. By this time, Patricia Talbott apparently had stopped making the excuse "That's just Bill" when he did something wrong and had started shouting at her son when she was displeased, Seat told Scharf. Talbott would shout back, and Seat would wish he could be anywhere else. He understood then why Talbott liked hanging out at his aunt Marilyn's place so much—he felt more accepted by that family than his own. Seat said he even observed Bill Sr. trying to slap or swat his son when he walked by, but with his limited range of motion, the ineffectual swings never connected.

Scharf started to notice something during the interview. Seat would pause for a second and shift his eyes to the side, as if debating whether to bring something up. Then his expression would clear and he'd pick up where he left off as if nothing had happened.

Seat recalled Talbott saying he had a horrible home life with his family, which is why he liked to stay with friends or his grandfather or out in the woods. He secretly liked it when his father would kick him out of the house and he had to live in the trailer in back. Talbott said it was the best way to avoid the bullying, by which he meant his family bullying him rather than the other way around. Seat felt bad for him, which is why he was so forgiving of what he called the less savory aspects of knowing Bill Talbott.

His habit, after a few beers, of bad-mouthing women and saying suggestive, vulgar things to them particularly annoyed Seat. He said that happened several times in the car while Seat was driving with a girlfriend and Talbott was in the back seat. Sometimes he would say things implying the girl was being paid to have sex with Seat. Other times he recalled Talbott would mumble, "I'm gonna fuck you. I should fuck you, too." When he'd get like that, Seat would tell him to shut up, but he rarely did. A shouting match would ensue, and Seat would finally slam on the brakes, pull the car over, and, rain or shine, day or night, make Talbott get out and walk home. Variations of this scene played out three times, according to Seat. After a day or two, a sober Talbott would stump up to him and mumble a gruff apology.

Again Seat paused, seeming on the verge of revealing something to the detectives that seemed to trouble him. Instead, Seat shifted topics to his and Bill's shared interest in photography. Scharf leaned forward a bit, eager to hear more on this topic.

Seat said Patricia Talbott had a small darkroom in the house that he and Talbott used sometimes, and that Bill had spent some of his motorcycle settlement money to buy a Minolta camera and telephoto lens, the same brand as Tanya's. He and Talbott would bring their cameras places, take a ton of pictures, then come back and develop them—rarely with Patricia's permission, because her son would never pay for any of the chemicals. Sometimes they would photograph at the speedway in nearby Monroe, or they'd hike some of the local wilderness trails and lake country. And one day, Seat recalled, Talbott led them on a hike—out by High Bridge.

Bingo, Scharf thought.

Seat pulled out a picture he had set aside to show him. "This just blows me away. I've had this forever. We took this maybe four years before this incident." This is the word Seat used instead of "murders."

It was a picture of the Monroe Honor Farm, the prison less than two miles from High Bridge, where the original detectives first looked for former inmates who might have killed Jay and Tanya and left the bodies nearby in a familiar location.

On the day of the hike, Seat drove but Talbott navigated, directing him to park under the bridge. Seat hadn't known how to get there without Talbott pointing the way. In addition to having the picture taken from that location as a reminder, it was a memorable event for another reason. They had been hiking into the back country and had to bypass barbed wire and other fences. And as they passed one, Seat felt an electric shock jolt him. He had been worrying that storm clouds were gathering for the last half hour and he shouted, "Bill, I got hit by lightning!" Seat was staring wildly around, but Talbott just laughed.

"It's an electrical fence, Mike," he said. "You grabbed it with your hand."

Then they both laughed, and Talbott helped him get around it safely.

Then Seat told the detectives about the house in Woodinville he shared with a roommate and old friend, Tim McPherson, who was a machinist at an aerospace parts company, Hirschler Manufacturing. They ran into Talbott in early 1987 and listened to him complain about being out of work and having to live in his sister Inga's basement. McPherson helped him get a job at Hirschler as a parts runner in March 1987, delivering aircraft fittings for contractors around Seattle, mostly aerospace behemoth Boeing and its many subcontractors. Once Talbott had a regular paycheck, he moved into the extra bedroom with Seat and McPherson, and they all shared the $475 monthly rent. Everyone was happy—except for the day they recall that Talbott tried to take apart and hack the cable box to get free premium channels and ended up frying the device. Then Talbott stopped showing up for work on time and lost the job in August 1987, and when he couldn't pay the rent, Talbott eventually left. Seat believed he lived between Inga's house and the trailer behind his parents' place after moving out.

McPherson would later tell Scharf that he used to drive the parts delivery route Talbott took over, and that one of the regular stops was a Seattle manufacturer just a block or two from Gensco, the heating supply shop where Jay and Tanya were headed and planned to park for the night.

Michael Seat at that point had become a gold mine and a key witness in the case: He had Talbott living in the Woodinville area around the time of the murders, close to High Bridge. He could testify that Talbott was familiar with the High Bridge area and had been there with him. And he had roomed with a witness who could show that Talbott was also very familiar with the place Jay and Tanya were headed, and where they could have crossed paths with their killer.

The conversation petered out at that point. Seat had that strange look on his face again, but he had fallen silent. Scharf was ready to wrap things up, pleased with the interview and feeling like he nailed down a valuable witness to help connect the disparate dots in this case. But he still sensed Seat was holding back something. So he asked his usual wrap-up question: "Is there anything else you can think of that we didn't ask you about Bill, anything that you had on your mind before we came here? From looking at the computer last night or talking to your aunt or anybody else, is there anything that we didn't cover?"

Scharf and his partner, Dunn, saw immediately that there was something else. The change in Seat was alarming. He paled and began to have trouble enunciating. "I'm struggling here with something, and I'm having a hard frigging time here."

He put his face in his hands and said something neither detective could make out.

"What is it?" Dunn asked as Seat continued to speak into his hands. "Okay, you're going to have to quit covering your mouth up."

"I'm sorry, sorry." Seat looked up. His eyes were watering.

"You're struggling with something that you want to tell us?" For a wild second, Scharf thought Seat was going to confess to being the killer—or at least the accomplice who the original case detective, Bart, always believed had been a part of the crime. The emotion Scharf perceived on Seat's face was one he had seen before when questioning people: guilt.

"Yeah, I . . . I . . . I have been think— I was thinking where in the hell was I going . . ."

Neither detective had any idea what he was talking about. But Seat finally managed to get the words out.

". . . when I saw the van."

At first, they thought he meant the pictures of the van he saw on the web. No, no, no, Seat said.

Scharf pulled a picture out of his file, the van with Jay and Lee Cook standing at its side. "This van right here?"

"Right, right, right."

"You thought you passed it someday?" Scharf asked.

"No. I seen it at Bill's parents' house. Sitting in the driveway. Early in the morning. I know I seen it there. God damn it, sticks in my mind. I been going through it all morning. Walking around the house, just thinking. I was going, Goddamn, I know I seen it sitting there. I know I seen it sitting there. . . . I just can't remember what I was doing that morning. I can't remember what date it was. It's been a lot of years, but I know I . . . I close my eyes, I can see it sitting there in the blackberries."

And gradually, with Seat struggling and near tears, the story came out, how he had seen Jay and Tanya's van in the driveway of the Talbott family home, how he distinctly remembered the hubcaps, how the picture in the news after Talbott's arrest had brought the memory to the surface, and how he had been torturing himself about where he was going that day. Was it to work? To the swap meet? Just for a pleasure ride?

Seat couldn't bring himself just then to confess the other reason he was stressed: his fear that he might have missed an opportunity to save Tanya. He just said he was struggling because he wanted a clear memory and to make sure he had it right.

Sensing he was still holding something back, and feeling that he had to make sure Michael Seat was just a witness and nothing more sinister, Scharf asked to take a cheek swab on the spot for DNA testing.

"Yeah, sure, that's fine." Seat didn't hesitate, which made Scharf feel better about his new star witness. Scharf knew he should try to repress his elation over Seat's startling revelations until he got that test result back from the crime lab. But he was having a hard time doing that. Michael Seat had just handed him something his predecessors on the case had searched for in vain for years: live, human, eyewitness testimony connecting Bill Talbott to a van whose last driver could only have been one person—Tanya and Jay's killer.

24

The End of the Perfect Crime

CeCe Moore had just solved another cold case, and she was in tears.

It was June 2018. Only a few weeks had passed since she identified William Earl Talbott. She felt triumphant with that solve, but now finding the killer of Christy Mirack, a sixth-grade teacher in Lancaster County, Pennsylvania, who died four days before Christmas in 1992, left her feeling bereft.

Yes, she found satisfaction in identifying the culprit, glad to bring answers, if not peace, to a family after a quarter century wondering who did this thing and why. Someone had grabbed Mirack as she opened her front door to go to work, carrying the Christmas gifts she had wrapped for her students, each with a personal note inside. The intruder punched her hard in the face, pummeled her back and front, then grabbed a wooden cutting board from the kitchen and bludgeoned her, breaking her jaw. He raped her and finally strangled her with her own sweater, then fled. Nobody in the neighborhood saw a thing.

Christy Mirack loved to teach; she had always known what she would do in life. It was obvious to everyone around her that she had the gift. Her favorite game as a child was to make up lesson plans, then cajole her younger siblings to come to class. So her principal knew Mirack was the least likely member of his faculty to be absent without calling in. After trying and failing to reach her by phone, he went to Mirack's house to check on her. He found the door ajar, and then found his star teacher dead, still in her coat and gloves but naked from the waist down, her face a bloody ruin, her students' presents scattered around her in a bright red and green arc.

The police in East Lampeter Township knocked on every door in the neighborhood, interviewed no fewer than fifteen hundred people, sampled the DNA of more than sixty potential suspects, created a website pleading for tips entitled WhoKilledChristyMirack.com . . . and came up empty. They assumed the killer was long gone, back to whatever hole he crawled out of. Detectives dreaded the day that the same unknown assailant's DNA showed up at another gruesome crime scene somewhere. But it never did.

It was the perfect crime. No clues. No witnesses. No hope.

CeCe Moore identified the killer in a matter of hours, more quickly, even, than she had found Talbott. The killer's half sister had uploaded a DNA profile to GEDmatch years before—one of the closest relatives and best head start on a cold case match list she had received yet.

But Moore wasn't quite prepared for what she found on the other end of the family tree this time. The killer had not crawled into a virtual hole, as William Talbott had done, seemingly cutting himself off from the world as if in training for a future in prison. Raymond Rowe, the man she linked through genetic genealogy to Mirack's murder, had been a prominent member of his community, hiding in plain sight all these years. He was all over social media, happy snapshots of him with his wife, proud-parent pictures of his children, celebrating special and everyday moments together like . . . like . . . There was no other way to put it, Moore thought: like a happy, normal, even enviable family, not much different from her life with her husband and teenaged son.

Homicide detectives grow used to this disconnect, this nesting of something evil inside seeming normalcy. Jim Scharf long ago made peace with the fact that there was no specific casting call for killers. Hell, he ended up arresting one child killer who was on his brother's bowling team. But CeCe Moore was new to this aspect of the job, and she found herself struggling with the fact that the man who destroyed the Mirack family also had a quite lovely family of his own. Now the members of that family, through no fault of their own, were about to be destroyed, too. And though she knew she wasn't the cause of this approaching tsunami of anguish—that would be Rowe's fault, and his alone—she still would be the vehicle for it. Rowe's wife, his fiancée at the time of the murder, would later say he came over the night he murdered Mirack, slept in her bed, celebrated Christmas with her a few days later, and talked about their upcoming wedding—even expressing concern for her safety after news of the murder broke. There's a killer on the loose, he said. You need to be careful.

Rowe had a thriving business as a disc jockey for hire, working under the name "DJ Freez." People loved DJ Freez, Moore saw on social media: one five-star review after another, dozens of them. His website claimed he deejayed events with Paris Hilton, the Eagles, and Sting, but the mainstay of his business was injecting a note of black-clad cool into local celebrations of all kinds. All Moore could think about was how his family was about to suffer, along with all these other people who had chosen him to be involved with their weddings, their kids' birthdays, their graduations, their bar mitzvahs, their special life-changing moments—now forever marred by the knowledge that they had partied and danced to the music of a murderer. Christy Mirack's own former students had sought out his services.

Moore made the ID in May 2018, less than two weeks after Talbott's arrest. As Jim Scharf had done with Talbott, the police in Lancaster County tailed DJ Freez, but it was easy getting what they needed. That same month, as he deejayed an event at, of all places, a local elementary school, police quietly scooped up a wad of his chewing gum and a water bottle he had sipped. The crime lab matched his DNA to the killer's. Within a few months, Rowe, by then fifty years old, confessed and pleaded guilty in court, offering an apology to the Mirack family, although he provided no explanation for why he targeted her. In exchange, he avoided a potential death penalty. His sentence of life without parole plus a consecutive term of sixty to one hundred years was designed to ensure he would die in prison.

MOORE, AT THIS point, lived and breathed murder every night, reading the most gruesome crime reports, then stalking killers online. There were dark circles under her eyes that no makeup could hide, and she started living in her sweats and pajama bottoms. She knew there was a personal and a family toll to living on the hunt for killers around the clock. And she absolutely could not stop.

If the police and public were surprised that one genealogist could solve two cold cases back-to-back in a matter of days—cases that had perplexed generations of detectives, the elite investigators in their respective departments—they hadn't seen anything yet. Moore cracked two more cases in June 2018.

In her third case, she matched DNA from the rape and murder of twelve-year-old Michella Welch, kidnapped on March 26, 1986, while riding her bike home from a park in Tacoma, Washington. Moore's genealogy, focused on a particular family-tree branch because of ethnicity and geographic clues from Parabon's Snapshot, led to a local sixty-six-year-old registered nurse. In the now-familiar tactic of trailing suspects until they drop something with their DNA on it, police used a crumpled paper napkin from the suspect's lunch to confirm the match between the murder and Gary Charles Hartman. They arrested him on June 20, 2018.

That same day, Moore identified the killer of Virginia Freeman, a forty-year-old real estate agent with a home to sell in College Station, Texas. On December 1, 1981, she left the office to meet a prospective buyer for the house, then never returned. The unknown "buyer" who met her there—the name he had given on the phone had been fake—bludgeoned, stabbed, and strangled her so violently that her neck broke. He left her body lying behind the house and drove off. But Freeman put up a fight before she died, and she scratched her killer. The skin and blood under her fingernails contained his DNA, useless in 1981, but preserved all these years until Moore used it to narrow her search down to a handful of possible relatives. One of them, James Otto Earhart, had been executed in 1999 for killing a nine-year-old girl—who died eight miles from the house where Freeman had been murdered. In her noncriminal genealogy work, Moore routinely asks people on the family tree she's researching to have their DNA tested to help fill in the blank branches, and Earhart's son agreed to do just that. His DNA results confirmed Moore's conclusion: his father had been the real estate agent's killer. To be doubly sure they could put this case to rest, investigators in Brazos, Texas, exhumed the executed man's body, and lab tests proved he was the unknown buyer who murdered Virginia Freeman.

Moore had used genetic genealogy to solve four cold cases in five weeks, two on the same day—cases that had been under investigation for a collective 126 years without results before she entered the fray. The cold case solve rate for the most accomplished police detective is something like one case every *year* or two.

She identified suspects in three more cases in July 2018. There was the 2016 murder of an eighty-one-year-old woman stabbed sixty-one times in her Woonsocket, Rhode Island, apartment.

Then there was the 1988 rape and murder of eight-year-old April Tinsley in Fort Wayne, Indiana, a notorious case in which the killer left taunting, menacing, and obscene notes claiming credit for the killing and threatening to strike again. Parabon had done a DNA mugshot of the killer two years earlier, but it was Moore's genealogy that ran down fifty-nine-year-old store clerk John D. Miller. When detectives knocked on his door and asked if he knew why they had come, he immediately volunteered: "April Tinsley."

The third case Moore took on that July was her first involvement with a recent crime still under active, initial investigation: the rape and robbery of a seventy-nine-year-old woman in St. George, Utah. Carla Brooks survived the attack but couldn't identify her assailant because she never got a good look at his face. She also told police she wanted her name made public so she could help raise awareness about sexual assault and the need to deal with the nation's scandalous backlog of unprocessed rape kits, which allows sexual predators to elude justice.

Moore got involved three weeks after the attack. The police had been unable to identify the culprit, and they feared he would strike again. He might even try to attack Brooks again, a detective on the case told Moore, as she refused to be forced from her home of thirty-seven years and the police couldn't watch her around the clock.

Maybe Carla Brooks was the fearless type, Moore thought, but the genealogist broke out into a sweat when she heard this. Someone was in imminent danger, either that septuagenarian or some other person on the rapist's hit list. Moore hadn't anticipated the cushion of time and history to be stripped away and replaced by urgency, although it made perfect sense now that it happened—an inevitability, really. Genetic genealogy was the new tool in a police detective's arsenal. Why should it be limited to cold cases? Moore stayed at her computer at least eighteen hours a day for the next three days straight to narrow the names on the match list down to four children of a particular married couple, just as she had done with William Talbott. Except this time there were four brothers instead of one. But it was enough. The brother responsible seemed almost relieved when the police caught up with him, admitting his crimes and later pleading guilty in court.

At the sentencing hearing, Carla Brooks spoke to the thirty-two-year-old who attacked her, Spencer Glen Monnett. Many in the courtroom were in tears that day. Even the judge's typically grim courtroom demeanor slipped when the victim told her attacker she forgave him.

"I fought like a tiger, but I ultimately lost. I was burned and you walked away. . . . I was seventy-nine years old, and I had to do a rape kit. . . .

"I'm glad I couldn't see your face that night so I can look at you today with no fear and move forward from this day on. You will leave here today with many high hurdles in front of you. I hope you will work hard because I want you to do well. I hope you get the treatment you need to get rid of the darkness that drove you to come into my home."

Monnett, who had been struggling with mental illness, porn addiction, and financial ruin, responded, "At that time, your fear and pain did not matter to me. My deeds were monstrous and redirected the course of your life. . . . I want you to know that I pray for your healing, and I could never apologize enough. Your pain does matter to me now."

Calling the hearing a "moving experience," Judge Jeffrey Wilcox said, "Carla Brooks, you are a magnanimous person. On a personal level, I am sorry for what you have had to go through. Mr. Monnett, I am sorry for you on a personal level. I believe what you've said."

The judge imposed a sentence of six years to life in prison, with the state parole board left to determine how long his imprisonment will last.

Through her long hours on her couch, guzzling coffee and staving off sleep as she built and tore down and rebuilt family trees on her laptop, Moore had made that moment possible. She had provided the crucial information the police needed to stop a dangerous person from hurting others, while also enabling the start of healing for both victim and victimizer. She wasn't just solving old mysteries and bringing answers to families who had been grieving for decades. Now she was also helping people deal with crimes in real time, and possibly preventing new crimes. That was true with Monnett, and also with April Tinsley's killer. Miller told police that he had made numerous other attempts to rape and kill children after April, but they would always fight or scream and, afraid of being caught, he always ran away. April was the only one who believed him when he promised not to hurt her if she kept quiet and didn't fight him. He had always been on the lookout for a "good girl" like her. If he had remained undetected, he might have found another victim.

BY THE END of September, Moore and Parabon had solved eleven cases. By the end of the year, the number reached thirty-two.

By September 2019, the count hit sixty-nine identifications involving seventy-five victims.

Moore wasn't the only genealogist working on cases in that first heady year and a half. Barbara Rae-Venter eventually came forward publicly as the leader of the team who identified the Golden State Killer, and she began to work on other cases, including one for Jim

Scharf. District attorney investigators in Sacramento trained by Rae-Venter did their own genetic genealogy research to catch a serial sexual assaulter known as the NorCal Rapist. The FBI's new genetic genealogy team assisted local police agencies in solving at least three cold cases around the country at the same time. And Colleen Fitzpatrick revisited a case she had tried to solve six years before through a Y-DNA search. This time she used a more powerful autosomal DNA profile uploaded to GEDmatch to solve the 1991 murder of high school student Sarah Yarborough, abducted and killed outside a school dance competition in the Seattle suburb of Federal Way. Her work led police to Patrick Leon Nicholas, a convicted sex offender who had avoided arrest in the past because his DNA had somehow never made it into the national criminal DNA database.

But the vast majority of solves involved CeCe Moore and the genealogy team she eventually assembled for Parabon to keep up with the caseload. She had become the de facto public face of genetic genealogy, with regular appearances in global print media and on news shows in the United States and Canada. She was the expert most often quoted on the subject, a fixture on shows such as *Dr. Phil*, with her own network reality TV series, *The Genetic Detective*, set to launch in 2020.

Headlines touted genetic genealogy as law enforcement's "major game changer" and "the future of crime fighting," while Parabon was dubbed "the Cold Case Factory."

It was only a matter of time before the backlash began.

25

The Nietzsche Dilemma

The extreme uneasiness many genetic genealogists felt about members of their profession becoming DNA supercops, using the databases and methods that they all had built and nurtured together, had very deep roots.

It began with Colleen Fitzpatrick's first work on the Sarah Yarborough case in 2012. Her Y-DNA search came up with a possible last name for the killer. And not just any last name. It was Fuller, a well-known American lineage whose ancestors arrived at Plymouth Rock in 1620. The headlines that followed were over the top: "Federal Way Murder Suspect Linked to Mayflower Pilgrim" and "Cops Hope Colonial Ties Reheat Cold Case."

There had been immediate pushback among some genetic genealogists who thought this was a gross misuse of archived family trees and databases that were not intended as a tool for jailing members of that family. Others were concerned that any sort of law-enforcement snooping through genetic databases would drive people away from having their DNA tested in the future, crippling the growth of genetic genealogy. Even innocent people don't want cops rummaging around in their private papers, much less their DNA, the critics said.

Others felt these concerns were overblown. Fitzpatrick's methods on the Yarborough case were standard genealogy practice and broke no laws or privacy rights. The same techniques had been used for years by many genealogists looking for clues to identify adoptees' biological families, to confirm marital infidelities, and for a host of other purposes that also were never intended by those who originally published their DNA profiles as a family-tree project. Why was it okay to end the anonymity of biological parents but wrong to help police identify a killer?

CeCe Moore neatly portrayed this divide—and where she stood—in a blog post at the time: "It comes down to this: If one of your loved ones was murdered and you believed that you could identify the guilty party using the same resources that we use for our hobby . . . wouldn't you?"

The discord faded in this first battle without any real resolution because, after all those spectacular headlines, the "Pilgrim connection" failed spectacularly to get results. There were an estimated 124,000 Fullers in the United States, and detectives hunting Yarborough's killer had never found reason to suspect any one of them. The limitations of Y-DNA matching made it a poor tool for crime solving. And as Fitzpatrick's later work on the case with autosomal DNA proved, the killer was not named Fuller.

The controversy returned in 2014 with a catastrophic misuse of Y-DNA. Police in Idaho Falls had done their own Y-DNA search for the killer of Angie Dodge, an eighteen-year-old who had been found raped and stabbed multiple times in her apartment in 1996. Investigators had come up with what they thought was a closely matching profile on a public hundred-thousand-profile Y-DNA database, which had been compiled by the nonprofit Sorenson Molecular Genealogy Foundation, then purchased by the consumer DNA company Ancestry. The database was treasured by genealogists, and Ancestry had kept it open and free. Idaho Falls police had not understood the limits of Y-DNA matching and falsely accused and terrorized an innocent New Orleans filmmaker. The headline in *Wired* was typical of the resulting coverage: "Your Relative's DNA Could Turn You into a Suspect: Using DNA for Criminal Investigations Is Great—Until It Turns an Innocent Person into a Suspect."

The worst fears of genetic genealogists concerned about police use of "their" databases were confirmed—not only by the negative publicity but by how Ancestry responded. The company, asserting that its data had been used in ways that were never intended, mothballed the Sorenson database with its prized hundred-thousand samples and associated genealogy pedigrees. No one—not police, not genealogists, and not researchers—would be able to use it any longer. It was considered a crushing loss to the genetic genealogy community and hardened attitudes against law-enforcement use of genetic genealogy. (Ironically, one of CeCe Moore's early solves was to find the real killer of Angie Dodge—and to help free another innocent man who had been wrongly convicted by the Idaho Falls authorities. This was the first-ever genetic genealogy exoneration.)

The latest backlash following the Golden State Killer and CeCe Moore's avalanche of solved cases started slowly, with the now-familiar low-level discontent among genealogists uneasy at the idea of police digging through their beloved GEDmatch database. Would this be Sorenson all over again?

While thrilled at the arrest of the Golden State Killer, lawyer and genealogist Judy G. Russell, blogging as the Legal Genealogist, voiced the concern of many of her colleagues by posting, "On the other hand, there is something deeply unsettling in the use of test results from what most test-takers consider a recreational use of their DNA by law enforcement in criminal investigations, when they had no idea the police could do such a thing."

Russell argued that people turn to genealogy to connect families—to discover new ties or renew broken ones. They uploaded their DNA to GEDmatch to make those connections easier. For the police to use that same DNA data without their knowledge to shatter families with criminal investigations is wrong, despite the good results, she wrote.

Russell and other critics pointed out that these investigations were done without search warrants, court orders, or regulatory oversight, and without informing and getting consent from the people who contributed their DNA to the databases in the first place. The Golden State Killer probe of GEDmatch was done in secret. Why? If police had announced they were after a dreaded serial killer and they believed searching GEDmatch would help them find answers and save lives, an overwhelming majority of users would likely have said yes, you have my permission. But they didn't ask. For some genealogists, this was yet another example of the Friedrich Nietzsche dilemma, the oft-quoted warning from *Beyond Good and Evil* in which the philosopher points out that those who battle monsters run the risk of becoming monsters themselves. Good ends, but bad means.

"Because some genealogical databases like GEDmatch are open to the public, it may be legal to search them for evidence in a criminal case," Russell concluded, "but what's *legal* isn't always what's *right*."

Criticisms grew more strident when, a year after the Golden State Killer arrest, a prosecutor disclosed that investigators on that case had obtained leads through creation of a dummy account used to covertly search a private consumer DNA database, MyHeritage. That search violated company restrictions and guarantees of customer privacy. Like its competitors Ancestry and 23andMe, MyHeritage does not allow law-enforcement searches. In the end, this didn't affect litigation of the Golden State Killer case itself. Seventy-five-year-old Joseph James

DeAngelo Jr. pleaded guilty to multiple murders and kidnappings in order to avoid the death penalty, receiving a life sentence without possibility of parole in August 2020. But distrust of genetic detective work was heightened by the deceptive use of civilian DNA data.

The problem for the genealogists who object to this use of technology was not catching criminals with genetic genealogy. They supported it, if done ethically. What they opposed was the use of people's DNA profiles without their consent. They saw this as unethical, no different than performing medical experiments on people without their knowledge or permission.

They saw two solutions if you ran a DNA database. You could do what the big companies in the space do—Ancestry and 23andMe, among others—and refuse to allow law enforcement, or anyone else, to upload a raw DNA data file, and go to court to fight any attempt to do so by search warrant as unconstitutional fishing expeditions. You could only get into their databases with a big vial of spit, which the companies process and upload themselves. No crime-scene sample is going to be able to fake that. On the other hand, if DNA database operators such as GEDmatch accept raw DNA data file uploads and want to let law enforcement do so with crime-scene files, the solution was just as simple: Tell your users exactly what law enforcement does and ask their permission. Get their informed consent. Those who approve can opt in to being in a section of the database open to matches with crime-scene evidence. Everyone else should be shielded from such scrutiny by default.

Just as this argument echoed ones from years earlier, the counterarguments remained the same, too. The genetic genealogy community had no qualms about using GEDmatch to completely upend people's lives to help adoptees and foundlings shatter the privacy of their biological families for a good cause. Why, then, oppose bringing killers to justice? Who exactly was being harmed?

From this side of the debate, the critics were nitpicking. Weren't their concerns about informed consent cured when GEDmatch announced within days of the Golden State Killer arrest that law enforcement was searching the database? Anyone who opposed this had been warned to make their GEDmatch DNA profiles private or remove them entirely. But this counterargument only further outraged privacy advocates, who said informed consent by definition can't be passive or assumed. It requires action. They liken it to a contract: you must sign on the dotted line or it's no contract at all.

The stakes were high. If the critics prevailed and their opt-everybody-out approach kicked in, the practical outcome would be that most of the million users in the database would automatically disappear from crime-scene searches. Cold case solving would become much more difficult or even impossible. Eventually people might get around to opting in, but many wouldn't—not necessarily because they opposed solving crimes, but simply because they weren't paying attention, didn't care either way, or had lost interest in genealogy and didn't visit the site. And a significant number of the uploads came from users who had since passed away. Those would disappear forever.

The two retirees who founded GEDmatch in 2010 were recreational genealogists who set out to build something for their fellow genealogy nerds. The service has always been free, with some premium features available for ten dollars a month. There were no employees, just the two owners and three volunteers, and the little service had chugged along in obscurity for nearly ten controversy-free years. Suddenly they were in the headlines, with police departments all over the country wanting to tap into their system. Half the genealogy world was displeased with them for doing too little to respond to all this, while the other half worried they'd do too much.

The managing owner, eighty-one-year-old Curtis Rogers, couldn't sleep for a week due to the worry and stress. His founding premise for

GEDmatch was to be open to all. Despite all the talk about genetic privacy Rogers kept hearing, as if intimate details about a person and their genome were being exposed, the only thing a GEDmatch search produced was a list of names of possible relatives, and their emails—if the user authorized it. That was it. And if cops wanted to use that information to find a serial killer, there was no way for Rogers to even know about it. This was an honor system. And, bottom line, he didn't want to keep the police out. He wanted dangerous criminals caught.

"The genie," as he put it, "was out of the bottle."

At first he thought being up front and informing users about what was going on would take care of the controversy. That was why he posted the announcement on his site about the Golden State Killer.

But a week after news broke of a second cold case solved through genetic genealogy—the murders of Tanya and Jay—Rogers, under pressure from the critics to do something more, changed the GEDmatch terms of service. They specifically authorized law enforcement to upload DNA and to search for matches connected to violent crime suspects, and to identify John and Jane Doe crime victims. But in trying to impose reasonable limits, Rogers miscalculated. Though violent crime is a category that includes numerous offenses (and attempted offenses), the new GEDmatch policy issued a very limited definition: only murder or sexual assault cases could be searched.

Nobody was opted out automatically, but users could always switch their listings to a privacy setting that would keep their personal information from appearing in other users' match lists. Or they could remove themselves entirely from GEDmatch if they were opposed to this new policy.

Rogers's solution wasn't what either side asked for, but everyone got something, calming the discord. The uneasy peace lasted for one year. And then it exploded anew.

The catalyst was an attack in a Mormon church in Centerville, Utah. A seventy-one-year-old woman was alone in a chapel, practicing the organ on a Saturday night in November after hours. The church building was locked. At around nine, she heard a pounding on the door to the building, but she ignored it. A half hour later, as she played, someone grabbed her from behind and choked her repeatedly into unconsciousness, then left her. Though the attacker may have believed she was dead, the woman survived, suffering two black eyes and large bruises and broken blood vessels in her neck.

Police found that a window to the chapel had been broken with a rock. The intruder had been cut while climbing through—there was blood on the sill and more blood on the doorknob from where the assailant escaped. DNA fingerprinting found no matches in the national criminal database.

That spring, with no leads and no suspects, the police turned to CeCe Moore. She refused to take the case because it fell outside the new terms of service. The woman survived what could have been a fatal attack, so it wasn't murder. If someone were caught, the charge would be aggravated assault, a serious, violent felony. But it was not a *sexual* assault. There were only two crimes permitted under the new terms of service, and this one didn't qualify.

The detectives in Centerville felt that made no sense. To them, this policy rewarded a would-be murderer for being bad at strangling. So they called Curtis Rogers and said just that. Did they have to wait until the attacker tried again and succeeded in killing? Because that was what might happen if they couldn't search GEDmatch for this strangler.

Rogers agreed to make an exception in this case and asked CeCe Moore to do the genetic genealogy workup as a personal favor to him. They were friends. He said he believed lives were at stake. Moore couldn't say no to him, and her findings led police to the juvenile attacker.

There was no mention of genetic genealogy when the arrest was announced. But it came out the following month, on the heels of revelations that consumer DNA testing pioneer FamilyTreeDNA, which had been eclipsed by newer companies but still had a million-person database, had been allowing law-enforcement searches for months without properly notifying customers. The GEDmatch exception pushed critics over the edge into open warfare.

The Legal Genealogist blogger, Judy Russell, announced that "GEDmatch can no longer be trusted." She had called GEDmatch a "DNA geek's dream site" in 2012. Now, she wrote, "That dream has turned into more of a nightmare. . . .

"GEDmatch made a decision, on their own, without consulting their users, to expose the data of those users to the police because the police made a compelling argument. Now it may very well be that the vast majority of those users would have agreed with that decision—if they had been asked. *But they weren't asked.* Somebody *else* made that decision *for* them. . . .

"*Bottom line*: nobody but nobody can give informed consent for someone else. Nobody but nobody has the right to make decisions for DNA testers other than the DNA testers."

Leah Larkin, a genetic genealogist who blogs as the DNA Geek, said she felt violated—to the point that she started rethinking her career choice. This was a person with a Ph.D. in biology and background in DNA research who had devoted thousands of hours to helping adoptees and other people of unknown parentage, including solving the mystery of the Hicks babies. These babies had been sold on the adoption black market for around eight hundred dollars apiece by a doctor in Georgia named Thomas Hicks between 1950 and 1965. The two-hundred-plus victims of this bizarre scam to steal and sell babies never knew their origins—and had only recently begun to trace their true heritage through genetic genealogy with the help of people like Larkin. And now she was thinking of quitting.

She criticized GEDmatch for betraying the trust of its customers and took Parabon and CeCe Moore to task for going along with it: "It's wrong. And GEDmatch has shown us that their contract with their users is meaningless. They should change their Terms of Service to read 'Anything goes. You're on your own.'"

Other influential genetic genealogists thought the criticism went too far. CeCe Moore complained that the critics were hairsplitting, that the terms of service never should have been limited to two crimes in the first place, and that the critics had essentially bullied GEDmatch into a no-win situation. Then they pounced when GEDmatch tried to deal with the illogic of allowing searches for murderers but not attempted murderers.

This was more a battle of influencers than ordinary users, but it generated headlines and rancor. It seemed that most ordinary GEDmatch users weren't terribly upset about all this. The site was not inundated with complaints from subscribers. New users continued to sign up, with the site reaching its highest point ever, 1.25 million uploads, *after* the controversy erupted. Many seemed to think that, yes, technically they should have been asked in advance for permission for new kinds of criminal searches, but most would have approved anyway.

In an age when there are far worse attacks on privacy afoot, when whole databases are being breached by hackers accessing truly sensitive personal, financial, and governmental information that could ruin lives, bankrupt families, and threaten national security, this search to capture the person who savagely attacked a seventy-one-year-old church organist seemed to some people to be an odd place to draw the line. To the critics, it was the principle of informed consent and the fear of the slippery slope—that one exception to the rules could lead to far worse ones in the future that might damage users in tangible ways. But so far, no one had offered any evidence that a GEDmatch user had ever been personally harmed by a law-enforcement search, or that his or her private genetic data had ever been exposed to the police.

But the criticism from respected leaders in the genetic genealogy community had an effect, along with fears that GEDmatch might face a costly lawsuit down the line. Within days, GEDmatch changed its terms of service again, giving the toughest critics much of what they wanted. The site expanded the list of offenses that could be used to match the FBI's definition of violent crime, adding manslaughter, robbery, and aggravated assault. But the more dramatic move was to create a choice for new users at GEDmatch: you had to choose whether you wanted to opt in to being in the part of the database searchable by police or to be visible only to non-law-enforcement searches. Existing users were automatically opted out, which meant there could never again be a question about whether users had consented. They had to go to the site and change the settings for their profiles to be searchable by the police.

It also meant GEDmatch went from being the most powerful tool in the world for cracking cold cases to barely usable overnight.

CeCe Moore was devastated. She thought it a rash decision, made out of fear, and it meant once Parabon's one hundred profiles ran out, she would be hard-pressed to continue her work, as would every genealogist who played by the rules.

"People will die," she told one interviewer. "It sounds dramatic but it's actually true."

Police detectives or genealogists working with them could still evade the terms of service by pretending to be ordinary citizen users. This new policy would encourage unethical but essentially undetectable behavior, she predicted. Rule breakers would be rewarded. Rule followers, punished.

Curtis Rogers vowed to encourage users to opt in, and the numbers slowly rose from twenty thousand to fifty thousand, but it would take more than a year to top two hundred thousand, with 80 percent still opted out. Even then, the database for sanctioned police searches remained crippled.

The fallout from this angry conflict lingered. Professional relationships dating back to the birth of genetic genealogy were sundered. Friendships ended. Nowhere was this more obvious than at the annual 2019 Southern California Genealogy Jamboree in Burbank a month after the latest iteration of the controversy erupted. This normally festive conference and series of workshops, lectures, and storytelling sessions draws fifteen hundred to two thousand genealogy amateurs and professionals pursuing cutting-edge genealogical research and evangelizing the virtues of home DNA testing. Pitchmen preview the next new products in genetic genealogy, while the workshops range from the technical ("Working with Your Autosomal DNA Triangulated Groups") to the practical ("Convincing Family and Strangers to Test and Why") to the poignant ("DNA and Uncovered Secrets: Help and Support"). In live sessions, genealogists walk newcomers through their DNA test results on the big showroom floor.

But for all its usual trappings, the conference was oddly muted and tense. The presenters had been cautioned not to discuss the law-enforcement controversy. Many regulars, most notably CeCe Moore, stayed away.

Outsiders took note of the discord and conflict. Legislators across the country were proposing new laws to establish ground rules and limits on police use of genealogy databases to solve crimes. A Maryland lawmaker wanted to ban it outright. Privacy advocates were clamoring for protections.

Then there were lawyers representing people who had been snared through genetic genealogy. They were looking for legal strategies that could free their clients. One theory getting some traction early on was the idea that DNA was so different from any other form of evidence, that it contained so much private information about, literally, every fiber of a citizen's being, that normal search and seizure law should not apply. People shed DNA all the time, involuntarily. It's part of being human. You can't help it. So, this legal theory suggests, old principles that allow police to scoop up evidence without a warrant if a suspect throws it away should not apply to DNA. Sure, when you toss a crumpled-up document into the dumpster outside your apartment, or leave a soda can with your fingerprints on it at your restaurant table, there is no "expectation of privacy," to use the legal term for one of the factors that determines if a search and seizure is constitutional without a search warrant. But DNA is different: there *is* an expectation that your DNA will still remain private when you lick an envelope or wipe your lips with a napkin, according to this view.

There is hyperbole in these arguments. The only deep dives and unlocking of personal genetic and health information in our DNA under way for law-enforcement purposes at this point involved the sort of work Parabon did with crime-scene DNA, where there is no expectation of privacy. Searching the public GEDmatch for names and emails of people who may be related to a violent criminal is definitely not the same thing. But many advancing these arguments didn't understand this distinction or conveniently chose to ignore or conflate it.

If the courts ruled such a legal analysis should prevail, it could mean all the cases that confirm genealogy matches by grabbing someone's trash without a warrant might have to be reopened, reexamined, or even thrown out. Which would be pretty much all of them.

And if that legal theory became law, then what about the DNA profiles in GEDmatch and all the other consumer DNA databases? You share DNA with all your relatives. When your uncle uploads his own profile, he is, in effect, uploading your DNA. That's what makes genetic genealogy so powerful—the simple fact that, at some level, we are all related. Is your relative violating your privacy rights with each DNA test he or she takes? You have no say over the process. Shouldn't searching such a database require a warrant? Were detectives and genealogists stomping on everyone's constitutional rights by failing to do so?

There was no court precedent that applied directly to genetic genealogy. Everyone was waiting for the first case to come to trial, for genetic genealogy to be put on the witness stand. And there was only one case that had a firm trial date and a defendant raring to go, eager to test the case against him.

When it came to the future of genetic genealogy, all eyes were on Snohomish County, Washington, and William Earl Talbott II, the first case of its kind to come to trial.

26

The Detective Versus the Four Pillars

May 2018 to May 2019
Everett, Snohomish County, Washington

Two things can happen when the focus in a murder case shifts from hunting a killer to prepping for a trial: the case can grow stronger, or the holes can grow bigger.

As the start date for jury selection of June 11, 2019, neared, Jim Scharf worried about the holes.

A year ago, he had felt more confident. The solution of an old mystery, making that call to the Cook and Van Cuylenborg families with the answer they had ceased to expect, had been deeply satisfying. Then on the heels of that win, he had solved one of his oldest cold cases: the sexual assault and murder in 1972 of twenty-year-old Jody Loomis, last seen leaving home on her ten-speed to visit her stabled horse. Genetic genealogy led Scharf to the killer, a retired heavy-equipment operator named Terrence Miller. That milestone was followed by promising developments in his quest to finally identify Precious Jane Doe. Best of all, his case against Talbott, in that first month after the arrest at least, was on the upswing—and seemingly everything else with it.

He had come to work the day after the Talbott press conference, his horses fed, his diabetic pug Cheyenne responding to meds, his wife, Laura, feeling better than she had in weeks. There had been a festive atmosphere in the office after the Talbott arrest, and the gloomy museum walk from the parking garage seemed to take no time at all. Suddenly the Snohomish County Sheriff's Office had earned national recognition for policing on the cutting edge, at the forefront of the genetic genealogy wave—not the norm for a smallish agency used to living in the shadow of bigger West Coast newsmakers in Seattle, Portland, San Francisco, and Los Angeles. The forgotten cold case office, Scharf, and his small group of volunteers—a retired judge, a retired parole officer, a retired prosecutor, and a self-taught genealogist—were getting attention and department love they hadn't seen since the unit began.

Scharf assembled significant new evidence in the Talbott case during that heady first month. Now he had four major pillars to form the foundation of his case—twice the number he had on the day he arrested Talbott.

First there was CeCe Moore and Parabon, who would spin the gee-whiz tale of genetic genealogy and DNA analysis for the jurors. This would explain how a man who had never been on the radar of any criminal investigation could suddenly be outed as a killer thirty-one years after the crime. The part of the case also served the nonlegal purpose of being the most glitzy and least boring thing jurors would see during the trial. A veteran of thousands of hours of court proceedings, Scharf knew all too well that, unlike cinema depictions of murder trials, the real thing is frequently dry and tedious. The gods of drama and pathos are shoved to the back of the courtroom most days by the lords of bureaucracy, as satisfying the legal requirements of chain of custody, relevance, and the complex rules of hearsay rarely make for good storytelling. After a trial, Scharf liked to chat with jurors and get their feedback on what worked and what didn't, and they invariably complained that witnesses got cut off just as they were about to reach "the good parts." Scharf feels their pain—he is often one of those witnesses who can't say what he longs to say. CeCe Moore would provide welcome relief and a bit of science mixed with Hollywood to counter that trial reality.

The second pillar of his case was the most critical: DNA fingerprinting. A DNA specialist with the Washington State Patrol crime lab, Lisa Collins, would testify that Talbott's DNA matched the semen found during the autopsy and on the cuff of Tanya's pants found in the van. The probability of the match being correct, Collins would testify, was 180 quadrillion to one. This was quite a bit better than saying no one else on the planet could match the crime-scene DNA. A single quadrillion is a one followed by fifteen zeros—a million billion. You'd need the population of 123 million planet Earths to have a large enough population for it to be likely that two people shared that same DNA profile.

Scharf had those two pillars at the time of the arrest, and they were solid. But he needed more, and after the arrest, he got more.

The third pillar came when he pulled all the old fingerprint files from the original investigation and asked the state crime lab's print specialist to see if she could match Talbott with any of the prints in the Cook family's van that had not been definitively matched to Tanya and Jay in November 1987.

After arresting him, Scharf had personally taken Talbott through the arduous process of creating "major case prints," which capture the full hand, not just the fingertips. The palm, the lengths of each finger and thumb, even the joints and the sides are all inked and their impressions coaxed onto file cards.

There were very few prints found in that search that did not match Tanya's and Jay's, and investigators had long suspected the killer wore surgical gloves in the van. But someone other than the Canadian

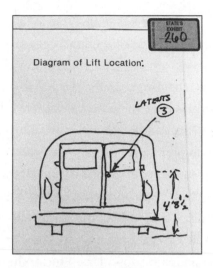

couple had left a very clear palm print at some point on the outside of the van, on the center edge of one of the back doors to the van's cargo area, just below the window.

Had someone unrelated to the case casually leaned against the van while it was parked somewhere, at a gas station or in the lot at Bellingham, which would mean the print was meaningless? Or had the killer gotten careless and rested his ungloved hand while befriending his victims, or abducting them, or while leaving their bodies behind? Scharf wanted to know. Three decades later, the fingerprint examiner had, after a rather lengthy delay that worried Scharf, reported that the palm print matched Talbott's. The identification was strong, the examiner told him—not a close call, but an obvious one. When Scharf saw the images, he agreed: even a layperson could see it when you scrutinized them side by side.

Scharf saw this as another major break in the case. The DNA evidence was powerful, but it only provided a direct link between Tanya and Talbott. The case was incomplete without a clear link to Jay, too, other than the fact that he and Tanya were traveling together. His body was found sixty-five miles away. His time of death may or may not have been the same as Tanya's—it was impossible to tell. If they were separated in death and time, the defense might argue that they had become separated in life as well. Jay was, after all, killed in an entirely different manner. The defense could argue in the absence of connecting evidence that someone else killed Jay, someone completely different in method and madness from Tanya's killer. Now that wouldn't fly; the palm print connected Talbott to the van Jay drove, a van owned by Jay's family, a van holding Jay's possessions, fingerprints, and DNA.

The fourth and final pillar in the case grew out of the interviews with Talbott's family. Along with all the stories about Talbott's behavior and outbursts of violence growing up, they had mentioned that Talbott had two roommates around the time of the murders: Michael Seat and Seat's friend, Tim McPherson.

This turned out to be the most useful thing the family members told the police, as none of their damaging recollections about his violent past, illuminating as they might be, were likely to see the light of day in Talbott's trial. Past bad behavior cannot be used as evidence in a trial for an unrelated crime. Only evidence that pertains directly to the murders of Jay and Tanya would be considered relevant. A prosecutor isn't allowed to persuade a jury that a defendant must be guilty because he did other bad things in the past. The only way that past misconduct could be brought up at the trial was if Talbott opened the door by taking the witness stand and presenting a good-character defense, claiming that he had never assaulted anyone in his life. Then, and only then, would the prosecution be allowed to rebut such claims with information from the relatives—assuming they'd actually testify about it. Scharf knew Talbott's defense lawyers would never let that happen, though it was always possible Talbott could insist on testifying and pry open the door himself. The thought cheered Scharf.

But the roommates were gold, and their testimony was completely usable and relevant. Through Seat and McPherson, Scharf could link Talbott to photography and high-end cameras, to High Bridge, to the Seattle neighborhood where Jay and Tanya were headed, and most incredibly (and hopefully not too incredibly) to Jay's van—sitting in Talbott's driveway, down the road from where Jay's body was found.

Just as importantly, Scharf had gotten back the results from the swabs he took of Michael Seat during their interview: to his profound relief, Seat's DNA had no hits in the criminal database and no link to Jay's and Tanya's deaths.

All in all, a month after Talbott's arrest, Scharf felt good about his case and its four pillars. This would not last.

THE TROUBLE STARTED during the next stage of trial prep, which consisted of rounding up all the old evidence gathered from the various crime scenes in two counties.

Scharf had already corrected the surprising failure to test for the presence of DNA on the flex ties. The lab reported back that he was right about them being possible sources of biological evidence: measurable amounts of DNA were detected on one of the plastic ties. But the quality of the DNA was too degraded by time and exposure to get a clear profile, as the ties had been stored all this time at room temperature. All they could say was there was a mixture of DNA present, possibly from three people. Current technology in the state lab couldn't separate that mixture into distinct profiles.

Linking Talbott's DNA to a flex tie would have been powerful additional evidence because the presence of the ties at both murder scenes suggests the same killer was responsible for both. The absence of DNA, however, created a hole for defense lawyers to point out. How could the killer's DNA not be on those flex ties if they really were part of the crime? And if they weren't, how could we be sure the same person killed both Tanya and Jay?

The rest of Scharf's trial prep consisted of making sure all reports and transcripts were made available to the prosecution and defense, and then contacting all the original investigators and witnesses in two countries and four counties. Many were retired (and a few had passed away), but a full complement would be needed to re-create the chain of events in 1987: the discoveries of the bodies, the autopsies, the preservation of the physical evidence and fingerprints, and the discovery of the van and the discarded items behind Essie's Tavern in Bellingham. Scharf had to locate, contact, and subpoena the grocery store clerks who could help chart Jay and Tanya's journey thirty years ago, plus the bartender at Essie's and the hunter who found Jay's body (the recycling scavenger who found Tanya had died). All of this took time and effort away from his other cold cases, but it was essential—not to prove Talbott's guilt, but to establish the elements of the crime and to forge the link between DNA found in 1987 and the DNA taken from Talbott in 2018. Every step had to be proven at trial, no matter how long it took.

Even as he assembled all the evidence, the holes he feared began to afflict his case. All four of his pillars came under clever and thorough attacks a year after Talbott's arrest.

The pressure on Michael Seat was the most dramatic. His testimony could be powerful evidence in prosecuting Talbott. If jurors believed that the van had been parked in the driveway of the Talbott family home, it was game over. There was no way to interpret its presence there without concluding William Earl Talbott II was a murderer. So Seat became the defense team's top target. They did not have to try very hard, Scharf realized. Even without them prodding him, Seat altered the story of seeing the van each time he talked about it. He injected different seasons, different years, different ideas of where he might have been driving. Sometimes he'd contradict himself in the same conversation. It would have been fine if he had just said, Look, half a lifetime has passed, and memory is like that. I also cannot remember what day I went to my first major league ball game or had my first kiss or got my first traffic ticket. Yet I can see those events very clearly in my mind's eye, and the van is the same thing. I saw it there, once, and it was never there again, and that's it. But Seat did not say that. He kept wracking his brain and trying to fill in the blanks, and each time the tale came out differently. Then the defense team interviewed Seat and got yet more variations. Next, because of his life-threatening heart condition, he was deposed under oath, which meant both prosecution and defense got to question him as if the trial were under way and he was on the witness stand. If he died before the trial, that deposition would be read aloud to the jury and it would count like any other witness's testimony.

The deposition was a disaster for the state's case. Seat was at first flustered by the aggressive defense cross-examination, then angered when he thought one of Talbott's lawyers laughed at him. Sputtering, he got up and walked out before the deposition ended. The defense lawyers left happy. They had all they needed to challenge Seat's credibility at the real trial. Just to be sure, they hired a well-known memory expert to testify that distant memories "recovered" in this way are unreliable and often false, even though the person doing the remembering is being sincere and telling what he or she thinks is the truth. "What Mr. Seat claims is not rationally possible," the defense lawyers said, previewing the expert's opinion in a legal brief. "That a person could recall a memory of a two-second viewing of a nondescript van parked in a driveway of a house that the person was driving past in the darkness thirty years ago defies rational acceptance."

Put that way, Scharf had to admit that it did sound ridiculous. But he had seen Seat struggle as he talked about this memory. He had watched him closely, looking for signs of deception or acting. For what it was worth, he believed Seat had seen the van in Talbott's driveway. But would anyone else? Would the jury? Or would they reject not just the van sighting but everything Seat had to say, including his other important recollections that connected Talbott to High Bridge and to living in that area at the time of the murders? Scharf had combed public records, bills, and payroll information, but he could find nothing that showed conclusively what Talbott was doing in November 1987 or where he lived. He had a degree of obscurity nearly impossible to maintain today if you owned a cell phone or used a credit card or surfed the internet—or submitted your DNA to check out your ancestry. Talbott's habit of listing his residence as a vacant property many miles away from where he lived and worked was effective then, and it served him well three decades later. The prosecution needed Seat.

It got worse. The palm print, the other pillar of the case Scharf found after Talbott's arrest, came under an unexpectedly vigorous attack. During the legal discovery phase—when the defense can question Scharf's witnesses—it came to light that Angela Hilliard, the state fingerprint expert on the case, had originally declared that Talbott was not the source of the palm print on the van door. Her opinion changed only after a senior examiner reviewed her findings and asked her to take another look. The second time around, Hilliard concluded she had made a mistake and declared Talbott a match. This was the cause of the delay that had made Scharf nervous.

The defense team was up in arms at this news, alleging that the fix was in—that this was an example of the higher-ups disliking an exoneration and exerting pressure to get a different result. Now Talbott had hired his own fingerprint expert, who was demanding all sorts of high-resolution photos of the prints as the defense mounted a full-blown challenge to the palm print evidence. Scharf knew that internal lab reviews were standard practice in every fingerprint case and were intended to catch errors no matter whom a reversal might help. But the optics here weren't great for his case, especially when the defense learned that in five similar reviews in the recent past, Hilliard had turned an exclusion of a suspect into a match every time.

Finally, in a classic legal judo move, the defense had undermined the first two pillars of the case against Talbott—genetic genealogy and DNA fingerprinting—by conceding them.

For genetic genealogy, the Talbott legal team had decided to drop—for now—any claim that the use of genetic genealogy to identify Talbott violated his constitutional rights, though the possibility remained that it might be raised on appeal if the trial ended in a conviction. His DNA was not in any genealogy database and so his "property"—in this case, his DNA profile—had not been searched, legally or otherwise. If anyone had a privacy claim, it was Talbott's two distant relatives whose matches led CeCe Moore to their and Talbott's doorsteps. But even if they did complain about invasion of privacy—which they had not done—that would have no bearing on Talbott's rights. If his DNA profile had been in GEDmatch and he was identified in that direct way, then he *might* have something to talk about, but that had not happened—not in Talbott's case, nor anyone else's.

This sounded like a victory for the prosecution and Scharf. By not challenging it, the defense would essentially be welcoming testimony about genetic genealogy before Talbott's judge and jury. As this was the first criminal case to come to trial anywhere in the world in which genetic genealogy played a role, this was a huge concession. It was not technically legal precedent—appeals courts, not trial courts, create that—but it did signal the justice system's legal acceptance of genetic genealogy for the first time. Just as that 1987 serial sexual assault trial in Florida had opened the floodgates for DNA fingerprint evidence in court, this move in little Snohomish County would influence other trials throughout the nation and the world.

But the effect of this concession did not feel like a win to Scharf.

The defense position meant there would be what was known as a "stipulation"—an agreement by both sides that genetic genealogy had led the police to investigate and ultimately arrest Talbott. Instead of live testimony, that short, dry stipulation would be explained to the jury. CeCe Moore and Parabon's Steve Armentrout, both on the prosecution witness list, would not appear to testify about how they cracked the case. Any additional information the jury needed on the topic could be handled briefly by having Scharf provide a little background testimony. And with that, the prosecution lost arguably its most compelling, timely, and fascinating testimony—exactly what the defense wanted, turning defeat into a tactical win.

They were also taking the same win-by-surrendering approach with the DNA fingerprinting evidence. The Electronic Frontier Foundation and the ACLU might be waging a long-term national battle to redefine how the law treats DNA—a fight that, if successful, could limit the use of genetic genealogy and the seizure of discarded DNA without a warrant. Someday that might help Talbott, but his lawyers decided to focus on the short-term battle that could win in court now. They did fight to exclude the DNA, but for bad bookkeeping, not bad science or an invasion of privacy. They argued that the chain of custody for the DNA swabs had been inadequate over the years. When the judge tossed out that argument, Talbott and his lawyers said fine. In that case, they were going to admit freely that it really was his DNA found in connection with Tanya—and then argue it didn't matter.

The presence of his semen didn't mean he raped or killed anyone, according to the defense. It just meant he had sex with Tanya. Consensual sex. They would claim there were no injuries to Tanya that would prove or even suggest she was sexually assaulted. Talbott's defense would be that he and Tanya could have hooked up while she was off on her own during that unaccounted for week in which both young people went missing. Then they parted company. And someone else killed them. It would be incumbent upon the prosecution to prove otherwise.

Beyond that, the defense claimed the whole investigation for decades was infected with tunnel vision, with investigators certain that whoever left his DNA behind had to be the killer. Never once did they consider the possibility that the killer was not the source of the DNA found at the crime scenes. All those suspects who were cleared because their DNA didn't match were cleared for the wrong reasons, according to the defense. The real killer could be among them. The real killer, according to Talbott's team, was still out there.

They were going to use Jim Scharf's thoroughness and ingenuity in obtaining all those DNA samples that cleared suspects as a weapon against him.

THERE WERE MYRIAD problems with the consensual defense, Scharf knew, starting with the rather obvious fact that Tanya was found shot to death wearing no pants or underwear, a pretty clear indicator that the sex and the killing were directly linked. Scharf considered this a desperation move, another variation of the tried-and-true SODDI (Some Other Dude Did It) defense, which in turn was a thinly veiled version of the old blame-the-victim defense trotted out in every sex crime since time immemorial.

But . . . he had to admit that the other dings to his case were weighing on him as the trial fast approached. If enough doubt about the other pillars of his case were raised, the jurors would seek some alternative theory to hang their acquittal on. That was the real purpose of the consensual sex and SODDI theory put out by the defense. It sickened him, because anyone who knew anything about eighteen-year-old Tanya Van Cuylenborg would feel this defense was preposterous. Did she really have to be victimized one more time? Did her family really have to listen to this? He'd have to prepare them.

It was one more thing to add to a very long list. Jim Scharf felt he was falling behind on everything that spring. His worries mounted, his police reports in other cases fell more and more overdue, and he felt stressed out about his ailing dogs and the health of his wife, Laura. All had taken a turn for the worse. Laura was hardly getting out of bed anymore, no matter how much he cajoled her. And he was not sleeping. When he made his second cold case bust with genetic genealogy in less than a year, solving the Jody Loomis murder, he had celebrated by staying home sick with the flu for three days.

With Loomis, there was none of the festive air that lifted his spirits when Talbott was arrested. Apparently genetic genealogy and solving cold cases was old news at his department these days. When he returned to work from being off sick, he had to sit in on the Talbott defense lawyer grilling his DNA expert, Lisa Collins. For the next four weeks he ping-ponged back and forth between Talbott trial prep, post-arrest interviews, and follow-up in the Jody Loomis case. A few days later he had to drive to Bellingham to sit in on the defense interview with Essie's old bartender. He drove back by way of Parson Creek Road and stopped to gaze at the spot where Tanya had been left and found.

Two weeks before the start of Talbott's trial, Scharf hit a low point. His daily log entry, normally a *Dragnet*-style "Just the facts, ma'am" recitation of daily activities, was unusually personal that day:

> *The office sent me to a class this morning on PTSD that told me I am all screwed up because I want to work hard to help people. They really don't care, they just want to be able to say they did their job by giving us training. They actually insinuated it's my*

fault I'm this way, so I need to talk to my coworkers to get healthy. They admit they know I am way behind on my work, but I also need to find a ½ hour each and every day to go outside and that will help fix me. They don't want to admit that it's really because they expect me to do the job of three people and they don't appreciate what I do. It's up to me to figure out how to deal with it and stay healthy.

Not long after hitting this low point, the families and friends of Jay Cook and Tanya Van Cuylenborg began to arrive from Canada to attend the trial they had been awaiting for half a lifetime. For Scharf, their arrival could not have been better timed.

One after another they warmly embraced him, clasped his hand, asked about his wife and his dogs, and expressed concern at how tired he looked. They thanked him so much and so extravagantly it began to embarrass him. But they wanted him to know they appreciated how he kept them in the loop over the past year, including them and protecting them at the same time.

"I don't know what we would do without you, Jim," Jay's sister Laura said.

And with that, Jim Scharf felt some of the tension in his back ease just a bit. This was why he did what he did, he told himself. This was his constituency. These courageous people, and Jay and Tanya.

"So are you ready for trial?" Tanya's brother, John, asked.

The cold case detective smiled. "I am now."

27

Is That It?

"Really? That's it?" Juror Number 2 exclaimed. "That's where they're going to leave us?"

The twelve jurors had all just trooped into their room. For the past ten days, the space had been their break room and their gathering room, their waiting room and their hideout. Now, at last, it was their deliberation room. The wrinkled paper sign, hastily Scotch-taped slightly crookedly to the outside of the room's faux wooden door, proclaimed it so:

DO NOT
ENTER
Jury
Deliberations in
Progress

The room was filled by the same big table, the same whiteboard, and the same not-quite-comfortable chairs as always. Yet now it all looked different. This was no longer a room for trial spectators. It was a room of twelve trial judges—twelve tired, stressed "triers of fact," as the juror's job description defines their role. With a half hour of work time left to them before the courthouse closed for the day, they had no idea where to begin or just how this process of trying the facts was supposed to work.

But there were nods of agreement and forced laughter at Number 2's rhetorical *That's it?* They were all feeling it. When the prosecution rested that morning, the jurors had exchanged wordless glances at all the loose ends left untied, the dots unconnected, the testimony they anticipated but now would never hear. Where was the "Aha!" moment? They were even more shocked when the defense rested mere minutes later without presenting any substantive evidence or testimony. Not a word from Bill Talbott or anyone who knew him—none of the good-character testimonials from his bail hearing. There was no defense expert or investigator to provide alternatives to the prosecution's views. The defense hadn't even cleared up the simple question of where Bill Talbott really lived at the time of the murders. Why leave that ambiguous? That aura of a man hiding from the world did him no good in the eyes of the jury.

Yes, they knew the defense did not have to prove anything, that the burden of proof rested entirely on the state. This right to silence had been hammered into them by the lawyers, the judge, and the ritual instructions they had to endure before they were finally allowed to rise from the jury box in a chorus of stiff joints and vertebrae popping and snapping. But the defense lawyers had created an expectation of more than mere silence with some very extravagant claims: of a consensual sexual encounter that ended without violence, of two different murderers who stumbled on Jay and Tanya days after the encounter with Talbott, and of a police investigation and crime-scene analysis tainted by bias, or worse. Which would certainly have been interesting, even case-deciding facts—if the defense had offered any credible evidence to back it up. But they had not. The defense lawyers had given them a story without evidence, while prosecutors had given them evidence without a story. They hadn't even given them a map.

The trial, the jurors realized, hadn't left them with answers so much as it handed them a puzzle. They would have to write their own story line and timeline, it seemed.

"Where are we going to start?" Juror Number 8 asked no one in particular. She had an idea but wanted to hear what others had to say first.

"The question of consent," one of the five men on the jury suggested. "That's a good question, obviously."

Number 8 exchanged looks with a fellow juror, another woman also in her early twenties. They both groaned, for they thought it was quite obviously not a good question. When the defense lawyer had described in her closing argument what she claimed was evidence that Tanya and Talbott had consensual sex, Number 8 wanted to throw up. The other juror agreed. But others on the jury thought the defense lawyer had a good point. They wanted to clear that hurdle before they could even think about convicting Talbott of the murders. So the first part of deliberations would have to consist of a lesson in basic human biology, as well as a serious debunking of the long-standing myths about consent that had been deployed with depressing success by generations of rapists on trial.

"This," Number 8 sighed, "is going to take a while."

IN THE LEAD-UP to trial, Jim Scharf spent hours with prosecutors working on just whom the jury would hear from and what they'd say. There were multiple interviews of witnesses—first by Scharf, then the prosecution team, and finally the interviews requested by the defense lawyers. There are no surprise witnesses in real trials. Scharf acted as a travel agent—arranging for witness transportation and subpoenas, rounding up all the documents connected to each witness, and making sure the right people came on the right day of trial. The prosecutors had to sort through hundreds of items in the evidence room, from those ubiquitous plastic flex ties, to the couple's last ferry ticket, to the twine and dog collars used to strangle Jay, figuring out not only what to introduce in court but in what order. The jurors would get to see and even handle all this stuff, and sometimes that up-close-and-personal testing and touching could make all the difference in a case.

So piece by piece, the team assembled the basic building blocks of a trial presentation, right down to which slides to use for the PowerPoint during opening and closing statements: Should they use the casual picture of Tanya with her dog, Tessa, or the dressy formal portrait of Tanya posing with her dad? They hunted everywhere for a picture of Jay and Tanya together, but they could find only the one rather poor snapshot Scharf had looked at so many times. Every detail mattered, and as the defense would be building their own counter-case with exactly the same level of care, part of the process was anticipating the other side's focus and spin.

Through it all, what the detective truly longed to discuss at length was his theory of the case. Scharf foresaw that jurors would want the state to connect the disparate, static pieces of evidence into a coherent narrative that explained what happened, when, and where. Scharf had thought this out in detail long before Talbott became a suspect, and he wanted to share with the jurors his interpretation of the evidence: How he believed the most likely scenario unfolded with Tanya and Jay boarding the ferry to Seattle and driving off at the other end alone. How they then probably parked in front of the furnace supply shop in downtown Seattle as planned. How once there, weary from the long drive, they probably stretched their legs outside the van, perhaps enjoying a last cigarette before bed. And how Bill Talbott, his murder kit of gloves, flex ties, gun, and bullets in hand, went looking for a likely victim not on the Olympic Peninsula he knew from camping with his grandfather, but in that same part of downtown, where he knew the dark, lonely streets so well from his frequent truck deliveries in the neighborhood. Then, spotting the girl he wanted, he appeared out of the gloom with his pistol in hand. Wouldn't he have perceived the tall, strapping Jay as the greatest physical threat of the two, and so ordered Tanya at gunpoint to bind him? Wouldn't he then either force her to drive or tie her up and slip behind the wheel himself? The evidence suggests a journey north from Seattle with three known stops, the first being High Bridge in Snohomish County. That would have been the first logical place to go, that lonely rural spot just down the road from Talbott's family home where Jay became the first to die, brutally but quietly, keeping Tanya in the dark. Next, a quick stop at the Talbott house, either to drop something off or pick something up, then northward to Skagit County, where Tanya was raped, shot in the head, and her body left in another lonely woodland spot. A short trip to the nearby city of Bellingham completed the crime, where Talbott dumped the incriminating evidence—ditching his murder kit and both of his victims' possessions behind a dive bar, after parking the van in a lot down the street. A bus back home was just steps away.

After bringing his idea up in several meetings, the lead prosecutor on the case, Matthew Baldock, finally shut Scharf down. "We're not going to tell the jury any of that," he said.

Then the prosecutor explained why to the flabbergasted detective.

THE JURY'S FIRST great debate was exactly what Talbott's lawyers hoped for. They had worked hard for it, beginning even before the jurors were sworn in. During voir dire, defense attorney Rachel Forde had laid the groundwork through this age-old process of questioning potential jurors. Her blunt query: "Who thinks a young woman on a road trip with her boyfriend would never have consensual sex with a stranger?" There followed a long and uncomfortable discussion, delving in Q&A format into how and why such a hypothetical tryst might occur. The prospective jurors had shifted in their chairs and spoken haltingly, some grudgingly, but almost every one of them eventually conceded such a hypothetical situation was possible under the right circumstances. And just like that, Talbott's defense had established an alternate narrative of a consensual encounter between victim and defendant before the jurors heard a single witness.

Forde had doubled down on this theme with her next question to the roomful of prospective jurors: "If a young woman is found murdered and is naked from the waist down, does that mean that she was sexually assaulted in the course of the murder?" Almost everyone thought that was by far the likeliest explanation. But in the responses, the defense team identified all who most readily brainstormed alternative explanations. Many of them became the prospective jurors most favored by the defense team. These leading hypothetical questions during jury selection are a lawyerly artform, with the goal of treading as close as possible to the line that separates permissible probing of attitudes from improper attempts to sneak in information about the case in the guise of a question. Forde was a master of pushing this particular envelope. By introducing these ideas at this early stage, she hoped to plant the seeds of doubt. Which, for a defense attorney, are the seeds of victory.

There was one hiccup: After listening to all this, one prospective juror said her own experiences and being a mother would make it hard not to be biased against someone accused of violence against women. The defense attorneys challenged her impartiality and demanded that she be dismissed. But after the woman promised to try her best to be unbiased, the judge denied the defense request, seating her on Talbott's jury. It was a small moment in a long and busy trial, but it would have unforeseen consequences.

Now the time for deliberations had come, and the jurors took up the question of whether the defense theory of consensual sex between victim and defendant held any water. Six of the seven women on the jury as well as several men made it very clear that they believed it most certainly did not hold water. Several pushed back hard against Forde's justification for it. The defense lawyer had tried to support her early hypotheticals by arguing that the presence of a mixture of Tanya's and Talbott's DNA found during the autopsy, combined with Tanya's lack of physical injuries (other than the gunshot wound), amounted to physical evidence of two willing, sexually aroused partners. This, Forde insisted, was "indicative of a consensual sexual encounter."

Juror Number 8 was having none of it. This was a classic rapist excuse, she said, that confused the human body's involuntary reaction to physical stimulus with the mental state of sexual arousal and consent. It was just an update of myths rooted in the medieval belief that rape could not cause pregnancy—supposedly because women must be willing partners in the sexual contact in order for conception to occur. In other words, pregnancy then, and the presence of vaginal secretions now, disproved rape. Number 8 decided to shock the other jurors with a little plain talk.

"Okay, I've had a few encounters myself that I know I didn't enjoy," she said, recalling a few unpleasant, though consensual, experiences. "But according to the defense, I apparently enjoyed all of it and it was totally fine."

Several female jurors chuckled in solidarity at this. The male jurors stared at this office manager from Everett, Cheyanne Palmer, who came to court every day perfectly coiffed with a bow in her short blond hair, a young woman used to being underestimated—until her obvious maturity and forthright manner made her a leader in the jury room. She added, "It just doesn't work that way."

Other women on the jury concurred. They weren't alone. There had been muttering in the courtroom gallery when the lawyer made those claims, with several people getting up and leaving, and they were not just members of the Van Cuylenborg and Cook camps.

Research shows that a majority of rape victims have no genital injuries and just under half have no injuries elsewhere on their bodies, so their absence proves nothing. As for biological evidence of arousal, all sorts of body responses associated with pleasure can also be involuntary reactions to unwanted physical stimuli. People can have tears of joy as well as tears of sadness, trauma, and pain. They might smile or even laugh when they are nervous or upset, as well as when happy or amused. Ticklish people who enjoy being tickled and ticklish people who hate it both tend to have identical reactions: they convulse with laughter. Does that reaction mean the people who hate tickling actually enjoyed it? Or gave permission?

"Ms. Forde's point," Palmer told her fellow jurors, "I know for a fact isn't true."

This didn't completely resolve the question of consent for all the jurors. Forde had sown those early seeds well, and some felt obliged to keep this defense theory on the table until it could be conclusively ruled out. But for a majority, the tactic had backfired, and the dearth of evidence in support of consent pushed this question to the background for the whole group. They were ready to move on to the next big gap in their knowledge of what happened to Jay and Tanya: the timeline. There were six days unaccounted for between the day they were last seen alive at Ben's Deli on the Olympic Peninsula and when Tanya's body was found on Parson Creek Road.

If the defense was to be believed, almost anything could have happened almost anywhere during those six days. They seemed to want the jury to conclude that those unknown days amounted to reasonable doubt all on their own. Were Talbott's lawyers right? Were there just too many possibilities, too many unknowns to conclusively declare Talbott a double murderer?

Looking through the exhibits, the jurors searched for a map to help orient them to the places, times, and distances that mattered in the case. But no maps were mixed in with the hundreds of photos, bits, and pieces given to them. So they sent out a note, asking the judge for a map of the state. The judge and the prosecution were okay with it, but the defense attorneys objected. No map.

So Juror Number 2, Robert Martin, whose occupation was county planner but whose degree was in geography, sat down and began to draw a map of their own. They would connect the dots—literally—as they struggled to figure out how events might have unfolded.

THE PROSECUTOR WAS adamant: using Scharf's theory of the case to explain how the evidence fit together was out. They would leave this speculative connecting of the dots and the creation of possible story lines to the jurors and to the defense. The prosecution would stick with what they could prove unequivocally, what the evidence clearly showed, and argue that those facts alone were more than enough to convict Talbott—even if it meant admitting that there were many questions they just couldn't answer.

Think about it, Baldock said. If we give them a theory to consider, they're going to spend all their time trying to poke holes in it, because it's their job to do so. They'll do the same thing with the defense theory. We know what that story will be—just look at the questions they asked during the interview with Jay's sister. They learned he liked to smoke pot sometimes, that he could be a little irresponsible at times. They'll run with that and say he could have left Tanya with the van to go find some weed, approached the wrong person, and ended up a murder victim. Or they'll say that while he left the van to make a purchase, Tanya and Talbott had the chance to meet and have their consensual sexual encounter. After they went their separate ways, the defense will say Tanya crossed paths with her killer.

Jurors would see through this as easily as Scharf, Baldock asserted, recognizing it as evidence-free speculation with holes big enough for Talbott to drive his truck through. But offering an opposing prosecution theory would communicate to the jury that these competing, unprovable stories must be essential to the case. What if jurors thought Scharf's theory had holes, too? What if they thought Scharf had it wrong, and the killer met the travelers on the Olympic Peninsula? Maybe they picked up a seemingly friendly Seattle-bound hitchhiker? Or what if the jurors thought the van never boarded the ferry in Bremerton? They could decide both the prosecution and defense theories were flawed and call that reasonable doubt where no doubt really existed, simply because they were led by both sides to focus on theories no one could prove.

Let the other side have the dubious, unprovable theory, Baldock said. And we will stick with the facts: We don't know how or where or when Talbott crossed paths with Tanya and Jay, we just know with absolute certainty that it happened somewhere after Ben's Deli. And that's enough. We can't prove who died first or exactly when, we just know Jay was gagged, bludgeoned, and strangled at some point. Then the killer left him lying in a spot Talbott knew well, just seven miles from the Talbott family home, with Jay's ID, car keys, and other personal effects missing and plastic flex ties strewn near the body. These are facts, not theories.

Likewise, the prosecutor said, we will freely admit we don't know how or when Tanya got to Skagit County or how briefly or how long she was captive. We just know where and how she ended up: she was left in a ditch sixty-five miles north of High Bridge; shot once in the head from behind; wearing no shoes, panties, or pants; and with Talbott's semen inside her. We know a shell casing that matched the bullet in her head was found nearby, as were more of the same flex ties. The defense story is what? That she decided to go for a walk like that in late November in a completely unfamiliar location in the Pacific Northwest and bumped into a killer other than Talbott? Isn't it self-evident that the person who most recently had sexual contact with Tanya had also killed her and dumped her half-naked body in the middle of nowhere, then drove off? The jurors don't even have to resolve the question of whether there was a rape or consensual sex in order to conclude that, either way, the person who left his DNA behind also shot and killed Tanya. Why introduce Scharf's unprovable theory, Baldock asked, to complicate facts that are by themselves both damning and unassailable?

Likewise, the prosecutor said, we honestly admit we have no idea where the van may have been all that time, or all the places it may have visited. We just know where it ended up: in a lot in Bellingham, fourteen miles from where Tanya's body was found, with her missing shoes, underwear, and pants in the back of the van, and with Talbott's DNA on those pants. We also know there were more flex ties inside, and Talbott's palm print on the van's rear door.

And finally, we know that behind Essie's Tavern, just a block from the van, somebody dumped Tanya's ID, Jay's keys, his dad's note, plastic gloves, a box of the same bullets that killed Tanya, and more flex ties. And we know a Greyhound bus depot sat next door, an easy, anonymous escape. Who needs a speculative theory when you've got facts like that?

Scharf nodded. He got it. He wasn't convinced, but the truth was that, after an arrest, the detective's role changed from the man in charge to an advisor, facilitator, and investigator under the direction of the lead prosecutor. Baldock and his cocounsel, Justin Harleman, would call the shots and have the final say moving forward.

Scharf respected Matt Baldock, but he was keenly aware they were very different people—different styles, different generations, different origins. Baldock was in his midforties, a good twenty years younger than Scharf, his gray and navy blue suits always crisp and tailored, his close-cropped haircuts and perfect posture giving him a bearing reminiscent of a Marine Corps officer. New acquaintances sometimes asked if he had experience in the armed services. He would smile and say no—his childhood dreams were filled with drums.

The son of two music teachers, Baldock went to college on a music scholarship with every intention of becoming a professional jazz drummer. He still plays to this day, but just for fun; pursuing it as a career had begun to sap all the joy from his music. So he switched his major to philosophy and English, graduated with a degree he belatedly realized had questionable utility in the job market, and took the advice of a professor who suggested law school next. After graduating and spending a year clerking for a Washington State Court of Appeals judge, where he witnessed close-up the mistakes lawyers make and appellate judges must try to fix, he took a job with the Snohomish County Prosecuting Attorney's Office (the equivalent of the district attorney's office in other states), and never looked back.

His specialty was quite different from the Talbott case, which, other than the unusual length of time between crime and trial, was a classic stranger homicide case. Baldock's main job was supervising his office's crimes against children unit, the same specialty Scharf took on at the sheriff's department when he was promoted from patrol. Witnessing the violence, misery, and damage visited upon children on a daily basis is not a job everyone can handle, at least not without having their souls crushed by the inhumanity they see close-up every time they report to work. You must compartmentalize to an extraordinary degree, to keep the horror from devastating your personal life, yet retain the empathy, patience, and commitment to work with damaged children while bringing their abusers to justice. Scharf and Baldock shared that calling, part of the small club of professionals who can survive, and even thrive, doing that kind of work long-term. And so Scharf put his trust in Baldock.

THE TIMELINE SEEMED quite clear at first to the jury, as the prosecution presented a witness-by-witness re-creation of the original investigation back in November 1987. The first deputies, detectives, and witnesses on the case, summoned out of retirement, clutched their old reports and notes and used them as memory aids. They walked the jury through the discovery of Tanya's and Jay's bodies, the painstaking efforts made to retrace their journey, and the largely fruitless pursuit of sightings and suspects.

Laura Baanstra was allowed to testify a bit about Jay and the kind of person he was—his budding romance with Tanya, his dream of being a marine biologist—as well as what she recalled of the trip to Seattle, including her solo performance of the family's traditional long wave goodbye, her last moments with her brother. And after that, she recalled the days that Jay was missing, obsessively watching the news then being called to the police department to learn that Tanya's body had been found. She couldn't go on, and the judge called a recess.

John Van Cuylenborg also took the stand, and several jurors found his testimony about his sister memorable for his smooth, lawyerly delivery as he recounted his father's desperate search for Tanya, never giving up in later years, hiring private detectives, traveling to New York to meet with an expert in serial killer profiling, resorting to psychic consultations despite a lifetime of resistance to anything with a

hint of the supernatural. The older brother's carefully controlled façade cracked toward the end, though, when he recalled accompanying
his father to that funeral home to look at the unidentified girl's body
found in Skagit County. He began sniffling and gulping air, even as
prosecutor Baldock asked the question he hated but had to ask, to sat-
isfy the rules of evidence and prove an essential element of the crime:
Was it Tanya?

"Unfortunately it was." At that heartbreaking understatement, he
looked for a moment as if he would cry, but he pushed it down and
continued on. It was that steely effort of self-control more than anything that led several jurors to tear up for him.

But empathy wasn't evidence, and the more the jurors heard, the
less they felt they knew.

The couple disappeared the night of their departure, November 18,
and then nothing more was known about their movements until a
scavenger found Tanya's body on November 24. Jurors were frustrated
that they heard no clear time of death—she could have died the night
she disappeared, or it could have been three days later, maybe more.
During deliberations, the jurors spent hours discussing what the killer
could have done with Jay and Tanya all that time. What happened during that week? Given that this was supposed to be an overnight errand,
and that Tanya was known for reliably phoning home to check in, the
jury didn't believe they were off on some voluntary escapade. But if the
killer had them that whole time, where could he have stored them or
imprisoned them? Did he bring them to his parents' house? Did he
have a hideout? Access to a warehouse? Or had the abduction, rape,
and murders all happened within a matter of hours, all on the couple's
first day missing?

This confusion was very good for the defense. In fact, they had
done their best to encourage it. But the prosecution had one big factor
in its favor right from the beginning of deliberations. No advocates of

outright innocence emerged on the jury. Not one juror ever said they believed the police had the wrong guy, or that they saw any shred of evidence pointing to another person as the killer. Every juror, from day one, felt Talbott probably was guilty.

But that didn't mean the jury was going to convict William Talbott, because the defense had its own advantage as deliberations commenced: many jurors had doubts. Whether, after all was said and done, those would be reasonable doubts remained very much up in the air. Certainly "probably guilty" was not good enough to convict anyone of jaywalking, much less murder. None of the jurors felt good about convicting another person and putting him in prison for the rest of his days. None of them felt their job was to avenge the victims or to bring closure to the family. As the woman who became presiding juror later recalled, "We just wanted to be sure. . . . Anyone who felt that Bill Talbott was responsible for these murders wanted to believe that it wasn't true. Nobody on that jury wanted to find someone guilty."

The jury elected Number 6 to preside—Laura Berner, a former paralegal turned tech-company employee at Nintendo and, more recently, Microsoft. She had inadvertently drawn attention to herself in the first few minutes of deliberations by nervously wiping clean the giant whiteboard at the front, leading to her nomination and election. She turned out to be a good choice. Whereas Cheyanne Palmer served as the jury firebrand, pithily distilling long discussions into their essentials, Berner became the patient consensus builder, with a knack for turning disputes into productive problem-solving moments.

Berner was among those who suggested that, once the question of consent was fully aired, they focus on determining who killed Tanya next. It seemed obvious to all that Jay's murder couldn't be considered independently of Tanya's, given that most of the evidence was tied to her, not him. The verdict, the jury felt, stood or fell on their decisions about Tanya. Only then could they turn to Jay's death.

With that in mind, Bob Martin's homemade map helped get them started. The simple visual aid—a pictorial representation of the testimony they heard—made clear the logical order of events. The geography was compelling: Starting point on the south side of the map in Seattle. Next stop: High Bridge in Snohomish County, about thirty miles northeast of the furnace supply shop downtown. Then sixty-five miles north to Parson Creek Road in Skagit County. And then to the final stop, fourteen miles northwest to downtown Bellingham. It was all laid out in a mostly straight line. Martin didn't know the exact mileage, but his map showed the general position of the key locations.

Jurors felt certain that the starting point of the crime was near the heater supply company in downtown Seattle, not on the Olympic Peninsula or the ferry. The jury thought Jay and Tanya made it across the Puget Sound, late but unscathed. Their evidence for this was the ferry ticket purchased shortly before departure time. These were Washington natives, and they all knew how their ferries worked, just as New Yorkers don't have to be told how the subway works. For a drive-on ferry, you purchase the ticket as you would pay on a toll road, then move forward into a waiting area until it's time to drive on board, which in this case likely would have been immediately. There would have been no good way or reason for Jay to reverse course and leave.

They also didn't buy the idea floated by Talbott's defense team that Tanya or Jay went out to party that night, perhaps separately. It had been a long, tiring day, with a late arrival due to wrong turns. Berner pointed out that the inventory police made of what was in the van showed that Tanya had packed none of the clothes or shoes she would want for a night out on the town, and that a young couple early in a relationship on their first big trip together would have been unlikely to go their separate ways in a foreign city at night. And though Tanya and Jay had been careful about keeping receipts to show their expenses— with Tanya even running back into the Hood Canal Grocery because she forgot the receipt—there was nothing to document a late-night excursion for drinks, food, or cigarettes.

(Neither Scharf, the prosecution, nor the jury knew about Tanya's last-minute invitation to her friend May Robson to join her on the trip to Seattle, which would have been further evidence that Tanya had no interest in striking out on her own to meet another guy. Robson was not called as a witness, and it never occurred to her that this painful recollection might have had any significance for the case. She disclosed it after the trial in a series of interviews with the author.)

That left a relatively small time frame for Talbott to cross paths with Jay and Tanya, only a matter of hours between the time they would have gone to sleep in the van and when they would have awakened to make the furnace purchase the next morning. Some jurors wondered if they stopped and asked the wrong person for directions to the furnace shop address, or where they might find something to eat or drink. Or perhaps the killer found them at the van. They could have been standing outside it, or the murderer might have banged on the door and made up some pretense—pretending to be hurt and in need of help. The exact scenario didn't matter; however it unfolded in that dark, deserted street, the gun was all the killer needed to take control and to force them to meet their fates.

Without ever hearing Scharf explain his theory of the case, the jurors had come up with the same basic scenario as the best way to piece together the evidence. Despite the jurors' complaint about the prosecution's failure to connect the dots, Baldock had been right. They did it on their own.

After the first full day of deliberations, Berner suggested it was time to "take our pulse" to see where everyone stood on the question of Tanya's murder. No one said, "I don't think he did it." A slim majority said they believed they could vote Talbott guilty of Tanya's murder. Five jurors said they thought Talbott killed her, but they could not say they believed him guilty beyond a reasonable doubt.

Had they ended deliberations that day, it would have been a hung jury. And no one in the room felt confident that they would ever reach an agreement.

28

For Me, That Closes the Window

Jim Scharf spent the busy days of trial on split duty.

Most of the time he sat next to the prosecutors at their table, their chief investigator, making sure the right witnesses were waiting outside to be sworn in, passing Post-its reminding prosecutors of some vital point or forgotten tidbit, and basically serving as a traffic cop keeping the case lurching forward, witness to witness.

His breaks were spent shepherding the Cook and Van Cuylenborg entourage, sitting with them out in the hallway between sessions, taking them to lunch, alerting them when testimony would be can't-miss good or too disturbing to bear. There were the autopsy photos and shots of Jay and Tanya where their bodies were found, blown up obscenely on the courtroom big screen. The jury needed to see those. The families did not. Without Scharf or one of the prosecutors warning them, they'd have no way to anticipate and then, once it started, no way to rise and escape a courtroom packed with media and spectators. When it all got to be too much, Scharf would herd Jay's and Tanya's loved ones onto one of the groaning, agonizingly slow courthouse elevators and bring them to a private conference room within the sheriff's department. There they could relax out of sight.

Toward the end of the prosecution's case, it was time for Scharf to testify. He had on his good charcoal suit and dark necktie, with a tie bar shaped like a little rifle, when he took the stand. First, he set the stage by explaining his work on the investigation as a cold case detective. He spoke of the years of dead-end leads and suspects he chased down, using every investigative technique he could think of or borrow from other cold case detectives since he took up his job in 2005. Genetic genealogy was just another shot in the dark, he said. It was little known to law enforcement when he had decided to try it.

Then Scharf provided the jury a short and sweet nonexpert's explanation of how genetic genealogy works—the sort of breezy overview that experts immersed in a discipline often struggle to provide. Next, he recounted the moment he received an email on his phone from Parabon CEO Steve Armentrout, asking that Scharf call him. He had been awaiting the results of the genetic genealogy testing of the DNA found on Tanya's pants, and he had expected a long list of possible suspects to check out, he told the jurors. But he got something else instead.

"It was on my day off. And I was out walking my dogs in the backyard when I placed the call."

Scharf paused a couple heartbeats then, taking a deep breath that seemed to catch in his throat. He was sitting in the old-school realwood witness box, the judge to his right, high on her bench, and the jury to his left, their full attention on him, close enough to hear the fabric of his gray suit move against his chair as he shifted position. In front of him sat the prosecutors' table, and to their right the defense, two lawyers seated with Talbott in the middle. The defendant glared at Scharf when he wasn't scrawling notes to his lawyer. Throughout his testimony, Scharf had pointedly looked at either Matt Baldock, who was asking the questions, or at the jury. He studiously avoided even glancing at Talbott.

Scharf resumed his testimony about the call to Parabon, saying, "He told me . . ." But his voice cracked, then failed as he teared up, his recall of the moment triggering the same response he'd had a year earlier.

He paused briefly again, then resumed in a voice at first tremulous. The jurors were riveted. Several blinked rapidly or put hands over their mouths as if staving off tears of their own.

"He told me that we found a match," Scharf said. "The person was a genetic match."

The prosecutor then asked the question for which even the least attentive listener in the courtroom already knew the answer, though everyone waited anxiously to hear it anyway, the tension reflected in the gallery's sudden quiet and in the silencing of the usual sound of seat fidgeting and muffled coughs.

"Did you get a name in that conversation?"

Now, at last, the detective looked directly at the man he had arrested three decades after the crime. "Yes. William Earl Talbott the second."

Then, with timing any movie director would applaud, the prosecutor said he would shift to another topic and this would be a good time to break. In the hallway a few minutes later, the Canadians engulfed Jim Scharf in a group hug.

Scharf had testified about how he discovered Talbott's name because the defense conceded the genetic genealogy evidence in order to downplay what might have otherwise been the flashy centerpiece of the case. But it turned out that despite the absence of star witness CeCe Moore—or perhaps because of it—Jim Scharf's testimony on the subject stood out, providing one of the most compelling moments of the trial.

Scharf's testimony certainly moved the jurors. Palmer, whose boyfriend once observed that nothing could make her cry, had to clench her jaw to stop the tears when Scharf recalled that moment.

In a trial where other police testimony tended to the methodical, even clinical, the contrast left a favorable impression with the jury. Palmer spoke of it afterward, saying it had been a relief to find that the lead investigator wasn't impervious but instead had been deeply invested in finding Tanya and Jay's killer, which suggested to her a detective less likely to make a mistake that could harm anyone. "Why would you not be emotional in that moment if you thought that a gruesome murder case was about to come to a close and the families would finally get some answers? You don't get into his line of work for yourself. Of course he was moved."

She had felt the need to articulate this in the jury room because defense attorney Forde had singled out Scharf for criticism. "When Detective Scharf heard the name Bill Talbott, he said the case was solved. He had the ultimate tunnel vision. His objectivity was so compromised that he cried."

This did not go over well with the jury. The job of defense lawyer is often thankless, and in a trial where neither the client nor anyone close to him would speak in his defense, there was little to do but lambaste the investigation. But this attack on the detective didn't just fail; it eroded the defense's credibility with jurors. Palmer used the word "scummy," and there was broad agreement in the jury room. They seemed to think Scharf's emotional moment on the stand did him credit, showing genuine compassion, not bias, in his work.

"The defense," Number 8 said, "did not read the room very well."

The other jurors agreed. And as more defense tactics started to backfire, the presiding juror felt the need to remind everyone, "We can't hold it against the defendant that the defense was horrible."

THE ONE PIECE of evidence the Talbott team attacked with unrelenting ferocity—Talbott's palm print found on the van door—also surprised the jury, and not in a good way.

Angela Hilliard, the state fingerprint expert, explained how she first believed the palm print from the van was not Talbott's, only to reconsider after an internal review suggested she needed to take another look. Hilliard realized then she had lined up the two prints incorrectly, which meant her microscopic point-by-point comparison of the van print with Talbott's known print compared different regions of the palm. This caused her to miss the match. But when she had them lined up correctly the second time around, she said the match was obvious.

Rachel Forde leapt on that assertion when it was her turn to question Hilliard, beginning by mocking the entire profession of fingerprint comparison as too primitive to be considered a skill, much less a forensic science. "Your analysis is a visual comparison, right?" Forde asked. "So it's just your eyes looking at one thing and seeing if it looks like another thing, right? Kind of what they teach you on *Sesame Street*, is that right?"

The jurors would later react to this the same way they did to Forde's attack on Scharf, perceiving it as both disrespectful and ineffective, unless the goal was to make the defense look foolish. "Then," Palmer said, "it succeeded very well."

Hilliard shrugged off the attack with a line that garnered chuckles in the courtroom. "I don't know. I didn't watch *Sesame Street*."

Next Forde led Hilliard through a lengthy review of a series of federal reports that criticized the fingerprinting field's poor system of tracking and detecting errors and its tendency to give extraneous case information to fingerprint examiners, which opens the door to unconscious bias in favor of matching a suspect to a print. Forde brought up a notorious example of this problem, in which multiple FBI analysts falsely matched an innocent Portland attorney to fingerprints found on a terrorist bomb in Madrid. And now, Forde suggested, Hilliard had done the same with Talbott: her boss had told her to look again at a decision that cleared Talbott, and that influenced her to declare the match her superiors wanted.

In fact, Forde said, in several other cases in which Hilliard took another look, she turned nonmatches into matches, but never the other way around.

That was true, Hilliard said, but only because she had overlooked a match the first time around, not because she had been improperly influenced or biased. "Unfortunately, it's because I'm human."

Also, she added, it was not her boss who did such reviews and kicked reports back for another look but a fellow senior examiner with no supervisory authority over her, and therefore no ability to pressure her to do anything. In her closing argument Forde continued to call the reviewer the boss anyway.

Still, this intense cross-examination over the failings of fingerprinting in general and in Talbott's case in particular provided a textbook setup for the defense to score major points by bringing in their own fingerprint witness. First bloody up the government's analyst, then bring in your own expert to say the print did not match Talbott, putting the prosecution on the defensive and throwing a great big reasonable doubt grenade into the jury room. The defense had hired a fingerprint expert to do just that.

But they did not call the expert to testify. Which could mean only one thing: their guy must have said the prints matched, too.

So all the jury ended up with was a really mean cross-examination that came with takeaways far worse for the defense than if they had just conceded the prints as they did the DNA. Those takeaways: there are national reports that say fingerprint examiners sometimes get it wrong and that systems are needed to detect those errors; that in the Talbott case, a mistake was made, a system of review designed to detect errors discovered it, and the error was corrected in a way that implicated Talbott; and the defense really disliked that result. But after much time and badgering of the state expert, the evidence against Talbott appeared stronger, not weaker.

The whole episode bewildered the jurors—not about the evidence, but about what the defense was trying to accomplish. The print seemed to prove only one thing—that Talbott had contact with the Cook family van, at least its exterior. But the DNA on Tanya and her pants was far more damaging, and the defense had conceded it was Talbott's without a fight. Why fight so hard to discredit the palm print then? In a consistent defense, it shouldn't matter any more than the DNA. To the jury, the defense had undermined its own case, because clearly they felt proof of Talbott's prints on the van did damage to his case.

To drive home the accuracy of Hilliard's final call on the palm print, Baldock had her walk the jury through slides of the two prints side by side on the big screen. As she highlighted areas on each print, and showed point by point where they matched, it became increasingly obvious to even a layperson that those were Talbott's prints. There had been no error, except initially when her orientation was off. At the beginning of her testimony, Hilliard had characterized it as "a little off," but when Baldock questioned her about this, she said the van print's positioning had been off by 180 degrees. That was more than a little off—180 degrees meant she had it upside down. Now it was obvious how she missed the match the first time: it had been as if she were putting a jigsaw puzzle piece in its place backward. None of the edges lined up.

Scharf, who had sat at the prosecution table silently fuming at how his witness and longtime colleague was being treated, abruptly forgot his anger and experienced a moment of revelation. The mistake sounded boneheaded, but it had been an easy one to make because the print left on the van had only the palm visible, no fingers. And in positioning it, Hilliard had assumed that if the fingers had been visible, they would have been pointing more or less skyward, the natural way someone would leave a print when pushing a rear van door closed from the outside. But, in reality, the print had been left by a hand with the fingers pointing down at the ground. As Scharf tried to visualize that, he felt sudden excitement, because he thought he understood what it meant, though he wanted the expert to confirm it.

After the break, he asked Hilliard how the print could have been left on the van positioned in such an unusual way, requiring a contortion of the arm to accomplish. Hilliard clearly hadn't thought about this before, and she paused a moment to consider. Then it hit her, Scharf saw, just as it had hit him.

"Well, this is really crime-scene reconstruction 101. He had to be inside the van. He was stepping down to the ground, and either braced himself or pushed off against the edge of the door frame."

That was it. The defense fussing about the print had helped them see what had been staring them in the face all along. Either when the killer was preparing to shoot Tanya, or getting rid of the body, or perhaps when he was leaving the van behind in Bellingham, he had jumped out of the van and, without thinking, left his print, the one time he touched anything without a glove on.

Scharf wondered, Was this the real reason the defense fought so hard to undermine the palm print? Because it did something not even the DNA could do definitively: put Talbott inside Jay Cook's van?

As the discussions in the jury room dwelled on the timeline and the doubts about what happened during those six lost days, the disparate pieces of evidence in the van crystallized into a sudden certainty and insight for Number 8, Cheyanne Palmer. She thought if she shared it, it just might bring to a close the Tanya phase of deliberations.

"Everything we need is in the evidence arrayed around her," Palmer told her fellow jurors. Talbott's DNA was found in autopsy swabs of Tanya, but it was not in her underwear found in the van. That, Palmer argued, shows us that after sexual contact with Talbott, she didn't have the time or the opportunity to put on her panties, as a person might normally do after consensual sex. When her pants were found in the van, Talbott's DNA wasn't inside them, either. Talbott's semen was found on the cuff. And Tanya was found naked from the waist down, except for her socks. Palmer asked, What does that tell us?

"To me, the evidence speaks to her not having time to put her pants on between the sexual encounter with Talbott and her death. She never had the chance to put on her underwear and pants. And that's how she was found. To me, that closes the window."

It was a breakthrough distillation that shrank the timeline and made hash of the defense's suggestion that Tanya could have been alive for days after meeting Talbott. Days without her pants or underwear? Days of activity, of going to the bathroom and cleaning herself, yet still with Talbott's DNA identified on swabs at her autopsy? "How does that work?" Palmer asked her fellow jurors.

As for what exactly happened next, maybe he shot her in the van, then drove her body to Skagit County so he could roll the body down the embankment and make a hasty getaway in less than a minute. The shell casing found in the brush could have come out of the van with the body. Or perhaps he shot her there by the side of the road. There was no way of knowing, and that wasn't important, Palmer argued. Either way, the evidence showed that Talbott was the last person to see Tanya alive. And he was the first person to see her dead.

Juror Number 2, Bob Martin, the mapmaker, is a spatial thinker. He had advocated putting a timeline up on the big whiteboard so they could see when events unfolded and how they interrelated. At first that timeline was huge and mostly empty, but now he agreed with Palmer. The timeline was mostly empty because everything happened at the front end. The span of time that the defense put out there as a stretch of days in which "anything could have happened" was, in Martin's words, a "red herring."

Everything had to have happened quickly, he argued, certainly within twenty-four hours. Sheer logistics dictated this: How else could one person hold two able-bodied, young, athletic people prisoners? More than twenty-four hours and suddenly the killer has to be a custodian of people for days, people who have to pee, who have to eat and drink. As does their captor, who would also be at risk of falling asleep at some point. How does that work—unless the murders both occurred within a day of the abduction?

Palmer was right, he said. Forget the consent question. It didn't matter. The evidence tells us the gap in time between the sexual contact with Tanya and her death was minutes, if not seconds.

There were probably no two people who were more different in age, life experience, attitudes, and origins than Jurors 2 and 8, so for them to land in the same place through their own shrewd analyses was powerful and persuasive. By the end of three days of deliberation, eleven of twelve jurors agreed with them. One still had doubts, but they all agreed it was time for them to turn to Jay's murder.

This was the first exposure to the American justice system for Jay's and Tanya's families and friends, their first extended visit to Snohomish County, and the first time they heard the details of the crimes that had robbed them of so much. They were edgy and fragile, likely to startle if tapped on the shoulder or jump when a pocketed phone buzzed.

Some mornings, in the brief period when the courtroom was open but before the judge took the bench, they'd clutch their notes or their purses or each other's cold hands and watch as defense attorney Forde asked prosecutor Baldock to help with her client's necktie. It was a strange, boundary-busting moment that happened several times during the eight days of trial. Inmates cannot have neckties in lockup—they too easily become strangulation tools against others or nooses for self-harm. Some public defenders keep clip-on ties in their briefcases for just this purpose, but Forde had a regular necktie, and Baldock would gamely put the tie around his own neck because, like most men, it was the only perspective in which he could actually tie the knot. He would arrange the necktie perfectly but loosely, then take it off and hand it to Forde to drop over Talbott's head and tighten the slipknot. He did not want to actually touch Talbott, but he performed this service without complaint. Was it gamesmanship on the defense attorney's part, the Canadian observers wondered, or just a genuine request? They didn't know, and it grated on them either way, but it confirmed their high opinion of Baldock's professionalism ("the Mensch," one nicknamed him) and their less flattering views of the defense team and the defendant, whom they had nicknamed "the Evil Leprechaun" over drinks one night.

There were two other surprises in the case that shocked and worried the Canadians. Powerful evidence was being left out of sight of the jury—at least until the case was over—and they were afraid this might tip the scales to acquittal.

The first involved Michael Seat, the only person who could be presented to the jury as an eyewitness to the crime, because he recalled seeing the van in the driveway of the Talbott family home near High Bridge. Expecting that his testimony would be too rigorously challenged, both by the lawyers and by the defense memory expert, Baldock decided he would not ask Michael Seat about seeing the van. The prosecutor said his main reason wasn't fear of the challenge to Seat's credibility, but that he and Scharf both worried that the stress of a long and angry cross-examination might be too much for Seat's fragile heart to bear, and that it might end up killing the man.

Scharf was uneasy about this decision to undermine one of the pillars of his case, and the family was even more concerned, but there was nothing to be done about it. They appreciated him being there at all, and so they silently sat, watched Seat's truncated testimony, and hoped for the best.

In truth, whether the motive was altruistic or strategic, this was another highly effective legal judo move, this time by the prosecution, as the defense had banked on using the van-sighting testimony to tear apart Seat's credibility on all matters and, the defense lawyers hoped, ruin the prosecution's credibility as well. Baldock had robbed the defense of what it hoped would be their most dramatic moment of cross-examination in the trial. Instead, Seat had a much easier time of it, while still providing some crucial pieces of evidence for the case when it came to Jay's murder—connecting Talbott to High Bridge by testifying about his family home nearby, recounting Talbott's love of photography and high-end cameras, and sharing details about the hiking trip Talbott and Seat took together in the area around High Bridge.

The other surprise in the trial was the late arrival of new DNA evidence—obtained from one of the flex ties found in the van.

In the lead-up to the trial, the crime lab had not been able to generate a useful profile from the flex tie Scharf had sent them because the DNA on it was degraded and consisted of a complex mixture from three individuals. This posed a difficult, sometimes insurmountable problem for any crime lab. But the Washington State lab recently acquired a new type of probabilistic forensic software that used artificial intelligence to separate DNA mixtures and fill in the blank spots caused by degradation. The lab's DNA supervisor, Lisa Collins, had told Baldock about the software after the trial began, and he had asked her to run the analysis immediately, even though it would be highly irregular to try to introduce new evidence this late in the game. When the results came in and he informed the judge, he explained that he had felt legally obliged to do the test because if the results were favorable to the defense—if, for example, some other man's DNA was on the flex tie—it would be unfair (not to mention illegal) to keep that quiet.

The results from this new software were explosive and not at all favorable to the defense: Tanya's DNA was in the mix. Traces of an unknown other person, probably female, possibly some sort of accidental contamination, were also detected. And there was one male DNA profile present that had a 90 million to 1 probability of being William Talbott. The plastic flex tie with this mixture on it had been found in the back of the van, snagged on a pair of Tanya's panties.

The DNA didn't just connect Talbott to the flex ties as well as to Tanya; it also blew away any argument the defense might make that the flex ties were meaningless or coincidentally present. It was, perhaps, the most powerful evidence yet in the case. The only problem: Baldock could not use it.

Just trying to get this new DNA finding into evidence would require delaying the trial to hold extensive hearings on the scientific validity of the new software, risking the creation of massive appeals issues. Baldock made the call not to push the envelope and to forgo using this new evidence. But the downside was that the defense could not only act as if the evidence did not exist; they could go further than that: they legally could tell the jury that no evidence existed tying Talbott to the flex ties.

And they did exactly that.

. . .

THE TIES HAD puzzled the jurors at first, because they seemed like a poor tool for a murderer, and the jurors were initially inclined to discount them. They weren't strong. They didn't have the metal reinforced latches that the police versions had. But when they pulled them out of the evidence bags in the jury room and saw one set of them strung together like a lasso, they realized how they could have been used to bind Talbott's victims over their jackets or a blanket, restraining Jay or Tanya without leaving marks on their skin. Once that had been established, those ties linked Jay to his killer as sure as the palm print and the DNA. They quite literally connected the dots for the jurors, even without knowing one of the ties held Talbott's DNA. Baldock had made another good call.

The cigarette pack used as a gag threw them at first, too, because Michael Seat had testified that Talbott didn't smoke and hated when people smoked around him or in his car. But then they realized the cigarettes didn't have to be Talbott's—they could have been bought by Jay or Tanya, and their use as a weapon could have been an expression of Talbott's anger and resentment against the handsome young man who got the girl.

Despite these conclusions in favor of the prosecution, the jury spent many hours considering ways to make the defense theory believable. They discussed how Jay could have left Tanya alone in the van in downtown Seattle to go find a place to buy a pack of cigarettes and, for some reason jurors couldn't understand, he decided to walk around a dark and ominous neighborhood rather than drive around with Tanya safely inside the van. And then a carless Bill Talbott came along and, contrary to everything the jury had heard about her character and history and relationship with Jay, Tanya decided to have sex with this hulking stranger six years her senior. But that led to an impossible fork in the road for the jurors. What happened to Jay at that time? What are the odds in that interval that something bad happened to him that led to his murder, completely unrelated to Tanya's fate—and his body just happened to end up a few miles from Bill Talbott's family home?

It didn't make sense. The jury rejected it. They couldn't find a reasonable way to disconnect the fates of Tanya and Jay.

It didn't even matter that they never heard Seat's testimony about the van. Talbott's connections to High Bridge were one coincidence too many for the jury. In fact, in their final vision of how the crime unfolded, they assumed on their own that Talbott very likely had driven the van to the family home and stopped there after killing Jay, with Tanya still in the van.

That Thursday night, after three days of deliberation, the jurors were 11–1 to convict on Tanya's murder, and 9–3 in favor of convicting on Jay's. It seemed to the jurors then—as well as to the lawyers and observers—that deliberations would stretch into the following week.

The next day, shortly after lunch, they defied everyone's expectations, including their own. They had reached a decision.

THE COURTROOM WAS packed—it was standing room only for the verdict.

The clerk read the verdict forms, the seemingly endless recitation of ponderous language of the case caption, the charges and the statute, and then the answer to the questions posed to the jury: Guilty of the premeditated murder of Tanya Van Cuylenborg. Guilty of the premeditated murder of Jay Cook.

It was a clean sweep for the prosecution: the jurors also voted for special verdicts finding three aggravating circumstances—that more than one person was killed in a common scheme, that the crimes involved the use of firearms, and that they were committed in the course of rape, robbery, or kidnapping. The findings assured the maximum sentence.

When the clerk read aloud the first verdict, at the word "guilty," Talbott flinched and slumped. Then he gasped, "But I didn't do it."

As he had not testified, these were the only words the jury ever heard him say. Several looked stricken, but Scharf stared at Talbott with unmistakable satisfaction.

As the family and friends of Jay and Tanya hugged and wept in the courtroom, Talbott collapsed into his chair. He seemed almost catatonic. Several bailiffs were needed to put him in a wheelchair to bring him back to his holding cell.

In recorded calls from jail, he had confidently predicted to his best friend that he would be acquitted. They were planning to celebrate the Fourth of July together, for surely he would be released by then. Was his confidence a sham? Or did he really believe he could sit on his hands the entire trial, explain nothing, and still walk out a free man? Scharf didn't know, nor did he care.

The sentencing would be in a month, but there was only one outcome possible in Washington State, where the death penalty was no longer available: life in prison without possibility of parole.

Genetic genealogy had won its first murder trial.

29

That's Gonna Stick in My Head

Reaching a verdict had been disturbing enough for the jurors in *State of Washington v. William Earl Talbott II*. But they felt positively gut-punched by his gasping protests of innocence at the end, the first and last time they would hear him speak. They returned to the jury room for the final time, shaken and hushed.

Not that they regretted or second-guessed their verdicts. There was no sudden change of heart. They were even aware of the galling hypocrisy of the moment: a defendant who chose not to assert his innocence on the witness stand when it counted, under oath and faced with the sort of rigorous cross-examination his lawyers dealt out to everyone else with abandon. But his daily taciturn and glaring silence at the defense table had left them unprepared for this final emotional outburst. Was it real? Had he really expected, whether from self-delusion or arrogance, to walk out of court a free man? Or was this a calculated moment for the cameras and the record? Whatever it was, it left the jurors feeling that much worse about their verdict's impact, visible right in front of their faces, mere feet away from the defense table, watching a man crumple in seeming pain and disbelief over their decision. Some of them could not wait to flee the courthouse after that.

"That's gonna stick in my head for the rest of my life," Number 8, Cheyanne Palmer, said afterward, "whether or not it's just the whimpers of a fool that got caught. Which logic tells me that's what that was. It still was so, so hard to hear."

The jurors who did stick around, however, met in the jury room with the two prosecutors and Jim Scharf to chat about the case and how the jury reached their decision. The visitors looked with interest at the big timeline on the whiteboard, sensing the hours and days invested in its details. And then a chorus of jurors said they knew there were things they weren't allowed to hear. Can you tell us now?

They said yes. Baldock recounted his decision to hold back Michael Seat's sighting of the van because of his heart problem and the memory expert, as well as Seat's belief that Talbott had been living in the trailer behind the family home at the time of the murders.

More meaningful for the jurors was the hidden testimony about the DNA on the flex tie. Many of the jurors had struggled to make sense of the ties, finally agreeing that while they did seem to establish a common theme in both murders, the purpose, if any, the ties had served remained unclear to some. Had they known about the DNA on one of them, their connection to Talbott and the killings would have been a no-brainer, and the verdict might have come much sooner, they said.

Then Scharf finally got to tell them his theory of the case, and he saw smiles on many of the jurors' faces. They had reached many of the same conclusions on their own, they told him, even speculating that Talbott must have stopped at the house with the van.

"We did it right!" Palmer exclaimed. "We did it like the detectives."

After the sentencing, a disappointed Rachel Forde emailed a statement about the jury she helped select:

The jury made the wrong decision. An innocent man will now be sentenced to life while the real killer remains free. The mystery was never in the DNA, but rather in the motive and the murder weapons possessed by the real killer. It's unfortunate that the jury adopted the police tunnel vision when it came to the DNA. Every American should now be concerned that the mere presence of their DNA at a crime scene could now lead to a conviction for a crime they didn't commit.

The presiding juror, Laura Berner, thought Forde's suggestion that Talbott's DNA was a "mere presence" at the crime scene was laughable, epitomizing a defense built on "a red herring that was easy for us to spot." Her take on Forde's statement:

It is true that the strongest evidence was certainly the DNA. But there was also the palm print, and a lot of other evidence, too. We figured out how it all fit together, and once we did, it was impossible to pull it apart. We were twelve very different people, with different experiences and different perspectives, and we all got to the same place. So she should know better. And she should have done better. The reality is we're right. We made the right decision.

William Earl Talbott's thoughts on the verdict would come a month later.

. . .

AT THE SENTENCING hearing, another packed courtroom awaited. And this time Talbott spoke, which he could do on this occasion without being sworn in.

"The level of violence in this is something I cannot comprehend," he said. Present were the families of the victims but none of Talbott's siblings. His father came but said nothing about him, or to him. He just watched in silence.

"I've gone all my life as a very passive person," Talbott continued. "I've never raised my hand towards anybody."

Scharf gritted his teeth but willed himself not to shake his head, as several members of the Cook and Van Cuylenborg entourage were doing.

Talbott then railed at the court for keeping out testimony about other suspects and tips that investigators had ruled out long ago. He claimed the true killer still roamed free and someday the truth would come out and he would be vindicated.

"The state says that all the answers here have been answered. Far from it. . . ." Then he faced Scharf. "That's your fault! I'm sorry, but— this is just beyond my comprehension, being charged with something of this nature."

Then, somewhat hastily, he added, "I do have sympathy for the family."

Willem Van Cuylenborg could not speak at the trial, a moment he had longed for right up until his premature death at age sixty-two. Nor could Tanya's mother, Jean, speak. She lived in a care home by then, Alzheimer's disease having robbed her of memory, including the knowledge that her daughter had been dead for all these years. In less than a year, Jean Lorraine Van Cuylenborg would pass away at age

eighty-seven, and join her daughter and husband at the seaside cemetery on Vancouver Island. So Tanya's brother, John, spoke for them all
at the hearing, as well as for himself, asking that Talbott, whom he said
had already enjoyed decades of freedom he didn't deserve, should
never be allowed to walk free again:

*My parents were never the same after Tanya's murder. After a
few years they were able to put on a façade most days that they
were continuing on, but it did not take much to scratch through
their thin veneer to expose their raw pain, shock at the cruelty of
the world, and their lack of will to carry on.*

*Personally it took me a couple of years to return to any sense
of normalcy. . . . The thought of doing anything enjoyable, even
listening to music or watching a movie, felt like a betrayal of Tanya's memory. . . . Thoughts of Tanya's final hours alive continue
to haunt me, and continue to leave me with a sense of frustration
and failure of not being able to help her in her hour of greatest
need.*

Jay's older sister, Kelly, rose to speak, despite her shyness and dis-
taste for public speaking. She had to do this for Jay, she said, recalling
how the defense began the trial by describing Talbott as an unremark-
able man caught up by the errant wheels of justice. She disagreed.

"It seems to me the only remarkable thing he ever did . . . was to
take. And what he took were the lives of two very young people."

And finally there were Lee Cook's brief comments, read from her
penciled notes:

*December 16 was Jay's birthday. He would have been twenty-one.
I had a birthday present ready to give him, and I had a Christmas
present for him. When I finally went to deal with his room, some*

of us wanted a shirt or a sweater of his you could wear, or hold them to your nose and smell him. I still have that old black sweater in my dresser drawer. For years I have heard him running up the back steps to the kitchen door.

The defense lawyers asked for a new trial, thundering about lack of evidence and prosecutorial misconduct. Attorney Jon Scott accused Matt Baldock of improperly appealing to sympathy and emotion over the fine young people who were killed, encouraging jurors to exact vengeance rather than focus on the facts. This was a bold argument from a defense that had sought to put the victims on trial, with suggestions of everything from infidelity and promiscuity to illegal drug and alcohol abuse.

Scott also came down on Baldock for another form of misconduct called "shifting the burden"—accusing the prosecutor of trying to use Talbott's absolute right not to testify against him, thereby shifting the burden of proof to the defendant to provide evidence in the case. But the statement Scott singled out for criticism didn't refer to Talbott at all but merely commented on the lack of evidence to support the defense's closing argument: "Ms. Forde mentioned again and again in her closing argument this innocent explanation, this innocent alternative explanation for why Mr. Talbott's DNA, his semen, would be on Tanya. Where is it? What is it? Have you heard it?"

Scott also raised the defense of consent again. The prosecution's case, he said, boiled down to suggesting that, because he supposedly raped Tanya, he did all these other things. But that was not enough, Scott said. Talbott was innocent and the jury had been bamboozled. Finally, he also said the jurors had committed misconduct by drawing their own map.

Judge Linda Krese brushed all these arguments aside as highly colored versions of reality that ignored all other pieces of evidence in the case, viewing each one in a vacuum rather than considering their collective power. She found Baldock's argument to have been an appropriate rebuttal to claims made by the defense. As for the map, all the locations in the case had been described in detail in testimony. Making a map by using that information was no different than taking notes.

The defense anticipated nothing less. They did not expect to win the day. They were laying the groundwork for Talbott's appeal, which he was already bragging about in those recorded calls from the jail as virtually a sure thing.

Then the judge imposed the maximum sentence, as the law required. Two life-without-parole terms—to be served consecutively.

Unless he won an appeal, earned a new trial, and then won that, William Earl Talbott's sentence meant he would never leave prison. If there were to be a new trial, the prosecution would try to introduce the DNA evidence on the flex ties. Seat could testify about the van. Another witness Jim Scharf had found too late to call, a former coworker who saw Talbott with a gun, could contradict the defense claim that Talbott hated and avoided firearms.

The case, conceivably, would be stronger the second time around.

Would Talbott try to improve his odds by testifying at a second trial, sharing his story if he got another shot? So far, William Earl Talbott's account of what he did or didn't do in November 1987 is the one mystery in this case yet to be solved.

AFTER THE SENTENCING, there was a gathering in Everett at a large restaurant near the water. An area was set aside with long banquet tables for the special guests. The Cooks and the Van Cuylenborgs and their friends were there. So were Jim Scharf, Matt Baldock, and Justin Harleman. Witnesses from the trial stopped by. The family members hugged and toasted Scott Walker, the man who had stumbled on Jay's body on Thanksgiving Day, whose hunting dog's name was so similar to Tanya's beloved dog, Tessa. Walker had emerged from testifying at the trial and begun to weep—for Jay and Tanya and their families and himself, a release that had waited more than thirty years. Even Chelsea Rustad, one of the distant cousins of Bill Talbott whose DNA on file in GEDmatch had helped CeCe Moore crack the case, came and was welcomed, much to her relief.

Something strange and wonderful unfolded next, a mix of sadness and elation, tears and laughter, sorrow and relief, sometimes all at the same time. Jay's friend Doran Schiller told stories. Tanya's best friend, May Robson, couldn't stop hugging Jim Scharf, who beamed and demurred when offered a beer or a glass of wine or a shot of tequila, all of which were present in abundance. He does not drink alcohol.

"But I'll have a Pepsi," he said with a grin.

He couldn't stay late. Tomorrow he would need to get back to work preparing for court in the Jody Loomis murder. His next cold case would soon go to trial. But not his last.

Dec. 3/86

There is a place that I know of.
Where up-above there flys a dove.
She slows and then she settles down,
And there is peace upon the ground.
But this is brief, as soon she'll fly,
And all beneath her then will cry.
And slowly as it turns to dawn,
She parts her wings and then she's gone.

Tanya

Epilogue

Once a week, sometimes twice, more cold cases are cracked around the country with startling regularity. In Snohomish County, Jim Scharf followed Tanya and Jay's trial by solving three additional cold cases through genetic genealogy, with more in the works.

The next to come to trial after Talbott was Terrence Miller, the seventy-eight-year-old man Scharf had arrested for abducting and killing Jody Loomis in 1972, one of his oldest cold cases. The pandemic had delayed the trial until late October 2020. But, finally, on November 9, 2020, the jury began deliberating Miller's fate.

The Loomis case had been resurrected from cold case limbo when a reevaluation of all the evidence by Scharf turned up previously undiscovered trace DNA on Loomis's boot. Genetic genealogy then led him to Miller. The same undercover cops, following their Talbott playbook, tailed the suspect to the Tulalip Resort Casino and grabbed something from the trash with Miller's DNA on it: a paper coffee cup. The déjà vu made Scharf smile. After a brief time in jail, Miller was released on bail with a tracking bracelet on his ankle. The police had confiscated his gun collection, but Miller had one hidden. As his jury deliberated, he killed himself. The jurors weren't told, and three hours later, the judge allowed them to deliver their verdict: guilty.

Scharf was not pleased by Miller's suicide. Jody Loomis's life, ambitions, and promise died when she was barely twenty years old, while the man who killed her stayed free, enjoyed life, even bowled on a team with Scharf's brother for forty-eight years. As far as the detective was concerned, Miller took the easy way out.

Scharf, on the other hand, had no easy choices these days, as challenges of health and home taxed all his reserves of humor and equanimity. He lost his beloved pug Cheyenne. Harley, by then blind, was ailing and would pass away in early 2022. The pandemic and Laura's compromised immune system had kept him working at home, minimizing their contact with the outside world. It's true that he could do his work anywhere, but the camaraderie of the office and the sanctuary of his paper-strewn cold case operation were gone, along with his weekly lunch and brainstorming sessions with his volunteers.

Scharf took solace in his farm chores, in his horses, and in his Man Cave, surrounded by his restored vintage cars and display cases of antique auto parts, traffic signs, and the trophies he and Laura won at the car shows they once loved to attend. And there was a new pug, China, to raise and train and accompany Scharf on his morning chores. Through it all, he still worked his cold cases, but once again Scharf had begun thinking it was time to groom a successor and retire at last. Still, he was not quite ready.

Closing Tanya and Jay's case and making history with the first genetic genealogy trial was certainly the capstone to his long career. But there was one other case he needed to finish before he could consider retiring.

He wanted to give Precious Jane Doe her name back.

. . .

CeCe Moore, meanwhile, showed no signs of slowing down. Her total identifications of criminals and John and Jane Doe crime victims reached an astonishing 175 by September 2021, when she addressed the thirty-second annual meeting of the International Symposium on Human Identification. At an earlier gathering of this forum, Moore was rebuffed when she first suggested that genetic genealogy could be used to identify criminals when other means failed. Now she is a featured speaker, responsible for the lion's share of the more than two hundred total cases solved by genetic genealogy nationwide so far.

The GEDmatch database, crippled as a crime-solving tool by new privacy policies, was bought in December 2019 by Verogen, a San Diego–based forensic sequencing company. Verogen restored the old policy of automatically opting in new users to law-enforcement searching, but the number of DNA profiles open to police remains barely a third of the million-plus profiles Moore had access to in the Talbott case. However, by searching both GEDmatch and the private company FamilyTreeDNA, which is also open to police searches, Moore and other genetic genealogists have continued to crack cold cases.

The divisions between Moore and some of the other pioneers of genetic genealogy have eased as well, though she knows things will never be as they were at the beginning. Gone for good are the heady days of collaborating and winging it with citizen scientist colleagues as they invented something new. Genetic genealogy is in the process of becoming a formalized discipline, one that can be taught in universities, with professional standards, guidelines, and licensing. In 2020 the University of New Haven became the first school to offer a graduate certificate in forensic genetic genealogy. Others will follow.

Moore knew this transformation was necessary, yet she still

mourned the passing of genetic genealogy's seat-of-the-pants era that first sparked her passion for this new science of identification.

"It will always be about families, though, about finding answers for them, some type of resolution," she says. "That will always be meaningful and fulfilling."

Meanwhile, the FBI is leading a committee of experts in the field in an effort to create standards and guidelines for law-enforcement use of investigative genetic genealogy. This working group will also study whether genetic genealogy can be incorporated into, or alongside, CODIS. The possibility of such a move added impetus to legislative and court-based efforts to place limits on law-enforcement use of DNA.

Maryland became the first state to pass such a law. After an initial effort to ban all law-enforcement use of genetic genealogy failed—a proposal fueled as much by myth and misunderstanding as fact—the legislature sought input from all sides of the issue: genealogists, prosecutors, cops, forensics experts, defense attorneys, and privacy advocates. This led to a new law, adopted in spring 2021, that seeks to balance protection for privacy and individual rights with allowing police to continue using genetic genealogy to solve crimes, but with judicial oversight. Police have to exhaust conventional investigative methods and CODIS before seeking a judge's permission to use genetic genealogy, and then only in cases of actual or attempted murder, rape, or other felony sex crimes and offenses that pose "a substantial and ongoing threat to public safety or national security." This is a fairly wide net—many offenses could fall into that last category—but the judge, not the police, will have the final say.

The genealogy community's privacy concerns were addressed as well by Maryland's law, which states that police can search consumer DNA databases only where users have given informed consent. And if a genealogical investigation would be aided by a DNA sample from a non-suspect—as when hobbyist genealogists ask for distant relatives to give their DNA to help fill out a family tree—the police must ask for this person's consent or prove to a judge that covert means are needed to avoid undermining their investigation.

In all, Maryland's law has something for everyone, preserving what police most value about genetic genealogy while providing oversight and accountability, including reasonable privacy protections that are unlikely to impede future investigations. After a rocky start, Maryland has provided a model for the rest of the country to build on.

A short time later, Montana became the second state to pass an investigative genealogy law, this one simpler and less ambitious than Maryland's. Its main provision is to require police to obtain a warrant from a judge before performing a genetic genealogy investigation.

APART FROM ITS regulation of genetic genealogy, the Maryland law had another purpose: it sought to bring transparency where very little existed. Police agencies, prosecutors, and crime labs must report annually on all genetic genealogy cases in the state—numbers, costs, outcomes, and demographics of suspects, particularly their race. It is the first (and so far only) attempt in the country to systematically track police use of genetic genealogy.

It's also the first official recognition of a serious equity problem with genetic genealogy investigations: like so many other parts of the justice system, it suffers from major racial disparities.

These disparities flip the justice system's usual chronic overrepresentation of Black men as both crime victims and suspects. Almost all of the suspects identified by genetic genealogy so far have been white men. And though the most common crime victim in America is a young Black man, most of the victims in cases solved by genetic genealogy have been white women and girls. When it comes to genetic genealogy, Black communities are being underpoliced rather than overpoliced.

The race disparity of suspects identified by genetic genealogy is mostly explained by the fact that an overwhelming majority of the profiles in the big consumer DNA databases in the United States are from people of European ancestry. The gender reversal, with most of the victims being women in these cases, is explained by the types of crimes that most often leave behind the suspect's DNA: rape and other sexual offenses, crimes in which the vast majority of victims are women.

Why the solved genetic genealogy cases should be so dominated by *white* women crime victims, as opposed to a more evenly distributed percentage of all races and origins, is not so easily explained. The Maryland law's transparency provision is a start in figuring that out. A national reporting system would be better.

WILL THE JOB be done when every state has something like the Maryland law in place? In a word, no. It's little more than a warm-up act, even a distraction, because the wrong issue has dominated the conversation about privacy and genetic genealogy. Focusing on law-enforcement use of DNA databases as a major threat to privacy is like regulating matches in order to address the problem of rampant wildfires. Attention is being misplaced—or diverted from—much larger potential threats to privacy and democracy.

While we obsess on what the police are up to when ferreting out a few names and emails from *public* genetic databases, millions of Americans are blithely uploading their complete genomic information to largely unregulated *private* profit-making companies who monetize customers' precious, extremely valuable DNA in a multitude of ways, including highly lucrative biomedical research. And, rather incredibly, the DNA donors are paying these companies to do it.

More than forty million people had taken a consumer DNA test by the end of 2021. That's nearly double the number reached in 2018.

What the police can access in their searches is nothing compared to the vast information these millions of customers are giving to private companies. It may go out the door as just a tube of spit in the mail, but to these companies, your spit is liquid gold from which your most sensitive, private self and secrets can be extracted: Are you prone to heart disease? Cancer? Alzheimer's? Mental illness? Depression? Do you have children with more than one spouse? Are you adopted? Are you related to a criminal?

People are giving away the keys to stuff even they may not know about themselves to profit-making companies who answer only to their shareholders. And the information you turn over to these companies also informs them about your children and your parents and your other close relatives—everyone who shares your DNA. You might as well send them your diary, your checkbook ledger, and your tax returns. But all the critics want to talk about is what the police are going to do with those names and emails they extract while hunting for serial killers.

It would only take one Enron of DNA, in an otherwise respectable industry—or one well-placed database hack of companies whose vulnerability has already been demonstrated—to cause more damage than anything imagined by those who worry about cops using genetic genealogy. What would that data be worth to an insurance company looking to deny coverage? To companies looking to screen their potential hires? To lenders and underwriters who make millions for every fraction of 1 percent of risk they can avoid? What would sensitive private information be worth to political operatives, domestic and foreign spies, to those who would blackmail leaders or manipulate and game an election? And the DNA doesn't have to be from the person being coerced. Malefactors can get to them through a cousin. Or an aunt. Or a child.

It's painful when your credit card is hacked. But you can cancel it and get a new one. Once your genome is hacked, there's no undoing it. It's the only one you've got.

The future is promising, frightening, and murky. The ability to detect and profile trace DNA found on objects, in rooms, on clothes, and even in fingerprints is growing by leaps and bounds. Genetic surveillance in real time is no longer science fiction. DNA is the witness that never lies, never forgets. Using it to arrest predators who thought themselves beyond reach is an obvious good. But where else will this take us?

FINALITY IS ELUSIVE in the justice system, as the families of Tanya and Jay found out when yet another twist in the murder investigation materialized where none was expected. In late 2021, the justices of the Washington State Court of Appeals gave William Earl Talbott II the victory he craved. They overturned his murder conviction and granted him a new trial. This was not because of any problem with the facts of the case against him, the conduct of the prosecutors, or any of the other complaints Talbott's attorneys made after the jury found him guilty. The court of appeals found Talbott's trial had been tainted by one biased juror. It was the prospective juror whom the defense had asked the trial judge to remove, who ended up seated as Juror Number 5. During voir dire she had said that she might struggle to be impartial in this sort of case, though she vowed, "I could try."

Her words in court that day that the appeals court found so problematic:

I'm an emotional person as it is, and I try to be very, very logical and methodical in decisions I make in my life and, you know, trying to see both sides of everything. But like I said, if it's a case involving violence and women, it's just something that I've already experienced in my life, and I fear that I will always inherently have as a mother, so that's just the one thing that I probably couldn't get past.

But she also added:

Just to note, it's something I usually express with my husband, that there's always multiple sides to a story, and I'm a fact-based person, so I could tell you that I will give it my very best, should I end up being on the jury, to do that. I just wanted to point this out to you, in case, in how you make your determination, that's a factor, you know. I'm an emotional being, like all of us, so it's just—the potential is there.

The trial judge had found that, in total, her words did not show an inability to be impartial but a desire to be open about how she might react to evidence in the case. But the appeals court said no—trying to do her best was insufficient. The justices seemed to want a clear statement that she *would* be impartial no matter what.

This ruling was based solely on what was said before the trial started, not on what happened in the jury room as the trial ended. The great irony is that, once the deliberations began, Number 5's fellow jurors found her to be cautious in drawing conclusions, open about her struggle to be impartial, pointedly skeptical of the prosecution's case, and among the very last to cast her vote to convict.

When this opinion was handed down, the Snohomish County prosecutors, though they used more polite terms, essentially accused the appeals court of going rogue. They said the justices skated over the fact that the defense had multiple peremptory challenges in its pocket, which could have been used unilaterally to remove any juror Talbott didn't like, regardless of the trial judge's wishes. Yet the defense chose not to remove Juror Number 5. In fact, the defense was on record accepting that juror at the time of trial. The prosecutors also reasoned that all any prospective juror can credibly promise is to try to do their best, and that jurors who are aware of their own biases and are determined to overcome them are probably better choices for jury service than those who bluster that they have no biases at all.

The county prosecutors have appealed to the Washington State Supreme Court to intervene. The high court could reinstate Talbott's conviction. Or it could rule in Talbott's favor and grant him a new trial. Talbott remains behind bars during this process, which could take months. Or years.

Jim Scharf is taking no chances. He's already organizing his files, preparing for the battle ahead.

A YEAR AFTER Talbott's conviction, genetic genealogist Barbara Rae-Venter finally called the Snohomish County Sheriff's Office with the information Jim Scharf had been awaiting so long. Her tip led him to just the right birth records and family members. And Scharf confirmed Elizabeth Ann Elder, born November 3, 1959, in Oregon, was his Precious Jane Doe.

Scharf had given the girl her name back at last.

Adopted at age two, Scharf learned, she grew up as Elizabeth Ann Roberts, though she went by Lisa. She was seventeen when she died.

Her adoptive parents had reported her as a runaway to the Roseburg, Oregon, police department on July 25, 1977. Her name and description were entered into the National Crime Information Center database for runaways, though for reasons Scharf could never uncover, the Roseburg police mistakenly removed her from the database that same day. And so the long mystery of her identity began.

Lisa called home from Everett, Washington, a few weeks after she ran away, asking her parents for money. Her mom begged her to come home. Lisa said no, but she promised to think about it.

Her family wired her money to an Everett bank, but no one ever claimed the deposit. Her family never heard from Lisa again.

"It took almost forty-three years, hundreds of hours of investigative teamwork by many volunteers, and extensive advancements in DNA science to make this happen," Scharf later wrote in a group email to family and friends of Tanya and Jay, who had taken a keen interest in his other cold cases. "Elizabeth Roberts, who went by Lisa Roberts, finally has her name back. Her family now has some answers of what happened to her. I'm sorry to say they are not the answers they wanted or needed to hear."

There was a postscript to the case of Precious Jane Doe that unnerved yet deeply satisfied Detective Scharf. He learned that the man who killed her, David Marvin Roth, had died while hospitalized for brain cancer a few years earlier.

The date of his death was August 9—the precise anniversary of Precious Jane Doe's death.

She parts her wings and then she's gone.
—Tanya Van Cuylenborg

ACKNOWLEDGMENTS

The Forever Witness could not have been written without the generous help and insights of those who lived this story, beginning with Snohomish County Sheriff's Detective James Scharf, who shared his time, his story, and his deep knowledge of cold cases and their investigation. Many thanks also to his partner in crime, CeCe Moore, the busiest person I know, and her fellow genetic genealogist Chris Green, who kindly opened doors for me as I began work on this book.

I am deeply grateful to the families of Tanya and Jay, whose grace under pressure leaves me in awe: Gordon and Lee Cook, Laura and Gary Baanstra, Kelly Cook, John Van Cuylenborg, Van DeGoey, and Bob DeGoey. May Robson, Tanya's best friend, shared invaluable recollections and histories of key people and events on Vancouver Island leading up to November 1987, even when it was painful for her to do so.

I also wish to thank Snohomish County prosecutors Matt Baldock and Justin Harleman; Paula Armentrout, Steve Armentrout, Thom Shaw, and Ellen Greytak of Parabon NanoLabs; Superior Court Judge Linda

Krese, now retired, for her unflagging support of public and press access to the justice system; public defenders Jon Scott and Rachel Forde; Michael Seat; the jurors who agreed to share their recollections of deliberations in *State of Washington v. William Earl Talbott II*; and history specialist Lisa Labovitch of the Everett Public Library's main branch, who was an indispensable resource on local history and whose outstanding library was my haven while in Snohomish County.

I am beyond thankful for the support and collaboration of that dynamic duo of Dutton/Penguin Random House: my stellar editor, Stephen Morrow, and rising star Grace Layer. And to my dear friend, tireless advocate, and literary agent, Susan Ginsburg of Writers House, thank you for joining me on this ride!

Finally, I am so very lucky to have the love and support of my family. Thank you! Without your endless patience in the face of the irascible creature I become as deadlines loom, I couldn't have finished one book, much less sixteen.

NOTES

Forever Witness Sourcing:

The events, scenes, dialogue, and biographical information in *The Forever Witness* are drawn from multiple sources that include:

- The author's interviews with many of the people depicted in this book; with investigators and prosecutors on the Tanya Van Cuylenborg and Jay Cook homicide case; and with people who knew Tanya, Jay, or William Earl Talbott II
- Testimony, exhibits, and documents in *State of Washington v. William Earl Talbott II*
- Police reports, interview transcripts, and other documents from the Skagit County Sheriff's Office and Snohomish County Sheriff's Office
- Author interviews with CeCe Moore, interviews with Steve Armentrout and Paula Armentrout of Parabon NanoLabs, and records on cold cases worked by Moore and Parabon
- Statements made by principal characters in the media, on tape or in print

- Author interviews with experts in genetic genealogy and DNA forensics, and presentations by experts in those fields at the annual International Symposium on Human Identification and the annual Genealogy Jamboree

PROLOGUE: LITTERBUG

The coffee cup:
Saliva is prized by crime labs because it is easily obtained, is hard for criminals to avoid leaving behind wherever they eat or drink, and is as rich a source of complete human DNA as a blood sample. That means even minute quantities left behind on such objects as a coffee cup contain genomic information more than sufficient for identification (or exoneration) of a suspect.

DNA analysis examines the arrangement of the molecule's four base chemicals into patterns—roughly analogous to lines of computer software code. DNA contains a fantastic amount of code, given that it carries the blueprint for an entire human being. By way of comparison, fourteen copies of *War and Peace*, one of the longest novels ever written, would be comparable to about a million lines of computer code. The self-driving Mars Curiosity rover, which has been exploring the surface of the Red Planet since landing there in August 2012, runs on five million lines of computer code. As of 2020, Facebook ran off a whopping sixty-two million lines of code.

But that's just a drop in the molecular bucket when it comes to DNA: the human genome contains the equivalent of 3,300 *billion* lines of code. This is so much information that, even though a person's DNA is 99.99 percent identical to everyone else's, that one-tenth of 1 percent still allows for a tremendous amount of variation in our traits—and therefore many ways to use DNA to identify a single individual from every other person on the planet. A trace of spit on the rim of a coffee cup, therefore, is one of the most powerful witnesses to crime in the law-enforcement arsenal.

CHAPTER 3: WRONG TURN

Tanya and Jay's route:
Skagit County Sheriff's detectives painstakingly re-created Tanya and Jay's journey from Vancouver Island south through the Olympic Peninsula to the ferry in Bremerton, based upon items found in the van and interviews with witnesses along the route. This was documented in police reports and court testimony in the trial of William Talbott.

The Richard Hugo quote:
This is an excerpt from Richard Hugo, *31 Letters and 13 Dreams: Poems* (New York: W. W. Norton & Company, 1977).

CHAPTER 4: PATROL DEPUTY SCHARF

Dino Scarsella, June 6, 1986:
Scarsella worked at his family's Snohomish County business, Scarsella Hay and Feed, and was known to single-handedly toss 120-pound hay bales up to coworkers standing above him in a truck bed. He'd carry hundred-pound sacks of grain two at a time, one under each arm, and load them for hours. Scarsella's doctor would later tell Scharf that under normal conditions, the man had something on the order of three times the average man's strength. In a manic state, off his meds, his strength might seem as high as seven times normal.

That's how, after nearly strangling his own mother and wife, Scarsella so easily overpowered Scharf and his backup deputy, then chased them from the house, across the yard, and into the street. Each deputy managed to lock himself in a patrol car. They radioed for help while Scarsella punched huge dents into Scharf's car with his bare fists. Later, with five cops on the scene, they barely managed to tackle Scarsella to the ground, and even then he almost stood up with all five men, easily a thousand pounds of cop, hanging off him. But they finally managed to handcuff him, bound his feet to his hands, lay him on the back seat of Scharf's patrol car, and drive directly to the nearest hospital.

Scarsella was alive and breathing when they reached the emergency room, and, fearing another rampage, the officers helped the medical staff get him on a hospital gurney and tied him to it facedown. When the doctor arrived, he had the deputies untie and roll Scarsella over. Scharf was horrified to see the man's tongue hanging out, and someone blurted, "He looks dead." Attempts to revive him failed.

A coroner's inquest, then a federal court jury in Seattle—which considered the family's civil rights lawsuit and allegations of excessive force—both held Scharf and the other officers blameless in Scarsella's death and found they acted reasonably under the circumstances. A combination of factors was said to have contributed to his death: his massive exertions, his mania, the sudden cutting off of antipsychotic medication, and his restraints in the car and on the gurney left the huge man unable to absorb enough oxygen. None of those factors alone would have killed him, but in combination, it had been enough.

Michael Seat seeing the van:
This scene was based on the author's interview with Michael Seat, as well as his testimony in court and recorded statement to investigators.

CHAPTER 5: SEARCHING

Willem Van Cuylenborg's search:
This chapter is based on sheriff's department documents, court testimony, the author's interviews with Bob DeGoey, John Van Cuylenborg, May Robson, and members of the Cook family, and statements made by Bill Van Cuylenborg on television and in news reports.

CHAPTER 6: FERTILE GROUND FOR THE BOGEYMAN

Vonnie Stuth and Gary Addison Taylor:
On the day before Thanksgiving 1974, nineteen-year-old newlywed Vonnie Stuth vanished without a word from her home. Her frantic mother, Lola Linstad, knowing this was completely out of character for her daughter, joined her new son-in-law in reporting her missing to the police but was met with official disbelief. Linstad was told Vonnie likely was absent by choice. Perhaps the

stress of a new marriage was too much for her. Or maybe she just needed time by herself. Wait and see, then get back to us. Linstad knew better. Vonnie was like Tanya when it came to staying in touch. But the police wouldn't listen.

Six months later, Vonnie Stuth's body was found buried behind her killer's house in a rural area south of Seattle. Gary Taylor, a thirty-nine-year-old machinist who once lived next door to the couple, had been arrested in Texas for another crime and confessed to four unsolved murders, including Vonnie's.

Taylor had a history of violence against women dating back to his teenaged years, including assaults, rapes, and the random shooting of women on the streets of Michigan in the late 1950s, when he was dubbed in the press "the Phantom Sniper of Royal Oak." Hospitalized for mental illness after confessing to having an uncontrollable compulsion to brutalize women, Taylor escaped and fled the state. His absence was not reported or posted on the national criminal information system for fourteen months, long after the trail went cold and he had preyed on new victims. He eluded arrest for years and is suspected in twenty other unsolved murders. He was sentenced to life in prison in Washington.

Linstad believed if police had listened, Taylor could have been caught sooner—perhaps while Vonnie still lived and certainly before Taylor's next victims died. She went on to cofound one of the nation's first crime victim advocacy organizations, Families and Friends of Violent Crime Victims. The nonprofit still operates in Washington, now as Victim Support Services, counseling crime victims and their families and assisting them in their dealings with police and the courts. Linstad's organization, and similar groups that sprang up nationwide, gradually transformed the way the system dealt with crime victims—an evolution that was still under way in 1987.

The Green River Killer:
Gary Ridgway would finally be identified as the Green River Killer in 2001, when he was charged and later convicted of a total of forty-nine murders. He confessed to seventy-one, but those numbers did not include Tanya and Jay.

The Coin Shop Killer:
Charles Thurman Sinclair opened a coin shop in New Mexico in the 1970s and expanded his business to sell guns as well. When the shop burned down in

1985, he lost everything. He would eventually be linked to the murder of an antique-shop owner in Snohomish County in 1980; the killer took $80,000 in silver dollars. All the Coin Shop Killer's other murder-robberies took place after Sinclair's coin shop fire and continued through 1990, committed by a man who befriended coin shop owners in California, Washington, Missouri, Montana, Utah, and Indiana. Each time, he earned the owners' trust—while casing the security of their operations—then after days or weeks, shot each one in the head and robbed their stores of all precious coins and other valuables. Robert and Dagmar Linton were his first non–coin shop victims, robbed and murdered while vacationing on the Olympic Peninsula in 1986, traveling the same route as Jay and Tanya.

Linked to at least eleven murders and other crimes, Sinclair was finally identified through coins he had stolen and was arrested in Alaska. He died before he could be tried for any of his crimes, but stolen property found in storage linked him to a number of victims, including the Lintons.

Spokane Serial Killer:
Robert Lee Yates was finally caught in 2000 and convicted of the Spokane Serial Killer's crimes. He committed almost all of his offenses while serving as a U.S. Army pilot. After his arrest, he confessed to several murders the authorities did not intially link to him, including a twenty-three-year-old Seattle woman, Stacy Elizabeth Hawn, who was kidnapped and murdered in July 1988. Her body was found near Big Lake in rural Skagit County. He also confessed to killing a young couple in Walla Walla in 1975 and dumping their bodies outdoors.

CHAPTER 7: THE JANE DOE OF PARSON CREEK ROAD

Unsolved murders:
German Lopez, "There's a Nearly 40 Percent Chance You'll Get Away with Murder in America," *Vox*, September 24, 2018, https://www.vox.com/2018/9/24/17896034/murder-crime-clearance-fbi-report.

FBI data on criminal case clearances:
Uniform Crime Reporting Program, "2017 Crime in the United States," FBI, https://ucr.fbi.gov/crime-in-the-u.s/2017/crime-in-the-u.s.-2017/topic-pages /clearances.

CHAPTER 8: IT'S ALWAYS THE BOYFRIEND

Murders by romantic partners:
Olga Khazan, "Nearly Half of All Murdered Women Are Killed by Romantic Partners," *The Atlantic*, July 20, 2017, https://www.theatlantic.com/health /archive/2017/07/homicides-women/534306/. (Note: The headline is incorrect and should read "More Than Half.")

CHAPTER 10: NOWHERE MAN

Surprising statistics on falls and mortality rates:
"The 50% mortality for deceleration injuries sustained from free falls is four stories (48 ft. or 14.6 m), and falls from greater than 60 ft. (18.3 m) almost uniformly are lethal," according to G. Alizo et al., "Fall from Heights: Does Height Really Matter?" *European Journal of Trauma and Emergency Surgery* 44 (2018): 411–16, https://doi.org/10.1007/s00068-017-0799-1.

CHAPTER 11: BABY ALPHA BETA
AND THE FINDER OF LOST SOULS

The first crime solved by genetic genealogy:
Before Baby Alpha Beta, there had been a few earlier partial assists in criminal cases with more limited genetic genealogy methods, including a number of cases in which genealogists produced a list of possible last names for police to investigate further. This method never pointed out an individual, just a last name, often several, and rarely was productive. Moore's methods, however, didn't provide just a clue: she used DNA to go from zero information to identifying both daughter and mother, victim and perpetrator.

CHAPTER 12: SHE PARTS HER WINGS AND THEN SHE'S GONE

Unsolved Mysteries show:
The segment on Jay and Tanya was originally broadcast on October 25, 1989. For details on the episode, see https://unsolved.com/gallery/jay-cook-and-tanya-van-cuylenborg/.

CHAPTER 14: THE DNA BLUES

Why Alec Jeffreys's shortcut created a sensation:
Ferreting out a simple method of identifying individuals by their DNA stunned the scientific world. It previously had been assumed that this could only be accomplished by decoding most if not all of the entire DNA molecule. That massive undertaking became known as the Human Genome Project, a moonshot-level research effort to map the entire DNA molecule, and it was years away from launch in 1984. It posed such a dauntingly massive chemistry and computational project that the $3 billion effort wouldn't be completed until the turn of the century.

DNA and innocence:
DNA fingerprinting has played a profound role in freeing the innocent and revealing mistakes by eyewitnesses, misconduct by the authorities, or flawed forensic methods that lacked the strong science behind DNA-based identification. As of 2021, there have been more than 2,800 exonerations of the wrongfully convicted through the use of DNA evidence based on Jeffreys's discovery—cases where DNA showed the wrong person had been imprisoned.

The twist in how the Leicestershire killer was caught:
After Jeffreys had examined more than five thousand DNA samples from all the male villagers in the area, he found none matched the killer. Then one of the men tested came forward and admitted that Colin Pitchfork had persuaded him to help switch samples, substituting his for Pitchfork's. Once tipped off, Jeffreys was able to confirm the match. Pitchfork then confessed.

Misconceptions about Mitochondrial Eve:
This ancient common ancestor should not be confused with being the first human on the planet, as if she just popped into existence in biblical fashion. The prehistoric individual Allan Wilson sought was neither the first of her kind nor a modern human. What Wilson hoped to find was the one member of a tribe of archaic *sapiens* somewhere in the world whose line of descendants survived across the eons, emerged from glacial ages, and successfully competed with stronger, faster, and fiercer—but less intelligent—hunter species. The descendants of this particular individual did well and mixed with other groups, passing on their matriarch's genes into the future as the species evolved gradually over thousands of generations into what we now recognize as human. That's what a common ancestor is—not the first: just the one whose descendants managed to be fruitful and multiply. Her genes are the ones that survive within us today.

Roots mania:
Glenn Garvin, "Remake of *Roots* Doesn't Sugarcoat Slavery," *Reason*, May 27, 2016, https://reason.com/2016/05/27/remake-of-roots-doesnt-sugar coat-the-evi/.

Margot Hornblower, "Genealogy: Roots Mania," *Time*, April 19, 1999, http://content.time.com/time/subscriber/article/0,33009,990751,00.html.

CHAPTER 15: THE TOOL OF INCLUSION

Fertility clinic fraud:
For an excellent compendium of fertility scandals see Jill Sederstrom, "Shocking Scandals of Fertility Doctors Impregnating Women with Their Own Sperm, Just Like in 'Baby God,'" Oxygen *True Crime*, December 3, 2020, https://www.oxygen.com/true-crime-buzz/are-there-other-fertility-doctors -like-quincy-fortier-in-baby-god.

CHAPTER 16: PRECIOUS JANE DOE

The Lisa Project:
Lisa's kidnapper was thought by police to be named Robert Evans, but this was just one of his many aliases. Investigators eventually found his true identity to

be Terry Rasmussen, a lifelong criminal born in Denver in 1943. He was eventually linked to four other killings in New Hampshire known as the Bear Brook Murders, in which the skeletal remains of a woman and her three young daughters were found sealed in barrels. One of the children was Rasmussen's biological daughter. By the time he was linked to Lisa's kidnapping and the New Hampshire murders, Rasmussen had been dead for five years, stricken by cancer while serving a life sentence for killing and dismembering his wife.

CHAPTER 17: FACING A KILLER

How Parabon developed Snapshot:

The body types, ear and nose shapes, skin tone, and other physical traits of the ten thousand volunteers were compiled into a database, and the supercomputer searched for correlation between clusters of single nucleotide polymorphisms—SNPs—and the many different physical and familial traits observed in each of the DNA donors. Which SNPs were always present when a person had a snub nose and which could be tied to a wide or prominent one? Sloping, straight, or curved forehead? A receding or jutting chin? Which SNPs were always associated with hazel eyes, and which are present for brown or green or blue? Could they find the combination of SNPs that are present for male pattern baldness or an olive complexion or a freckly face? This was a massive task. There are seventy-seven different genes that code for eye color alone—so each facial feature and physical trait required a huge amount of data and posed a knotty and lengthy computational problem. But each additional data point also increased the predictive accuracy of the DNA mugshots.

The end result: Parabon scientists reported they could predict physical features and overall appearance by analyzing an unknown person's DNA and assign a probability of accuracy for each feature: a 98 percent likelihood of blue eyes or a 90 percent confidence that the suspect has fair skin, or 93 percent certainty that the suspect is a white man of northern European ancestry versus a less than 1 percent chance that the suspect is of Middle Eastern descent. Parabon called this process "phenotyping," although this is an old term originally defined as a catalogue of all the observable characteristics of an individual organism that, in sum, distinguished it from others. In this modern version, phenotyping means cataloguing the physical expression of a living being's genes.

CHAPTER 22: OH, THAT'S JUST BILL

Initial skepticism about the value of genetic genealogy:
Solving the Golden State Killer case was initially perceived as an aberration by both experts and the media, who concluded that genetic genealogy work in the case had been so complex, time-consuming, and resource intensive that it could not possibly be replicated soon or often.

"It is unlikely that the apparent success of the method in the Golden State Killer case will spur a rush to use genealogy databases to solve crimes," *The New York Times* reported right after the big arrest. The article quoted New York University professor Erin Murphy, an expert on police use of DNA and privacy, explaining why it would be so rare: "Using a database of this kind will generate an extraordinary number of leads, and running them all down using both nongenetic and genetic information requires a lot of police power. So I doubt it will be run-of-the-mill anytime soon." This proved to be a sensible yet completely mistaken interpretation of the paradigm shift that genetic genealogy represented. Gina Kolata and Heather Murphy, "The Golden State Killer Is Tracked Through a Thicket of DNA, and Experts Shudder," *New York Times*, April 27, 2018, https://www.nytimes.com/2018/04/27/health/dna-privacy-golden -state-killer-genealogy.

CHAPTER 24: THE END OF THE PERFECT CRIME

Carla Brooks's concern about the backlog of sexual assault investigations:
As many as two hundred thousand rape kits sit unopened in police storage around the country, shielding rapists from being found through the CODIS criminal database or through genetic genealogy investigation. Barbara Bradley Hagerty, "An Epidemic of Disbelief," *The Atlantic*, August 2019, https://www .theatlantic.com/magazine/archive/2019/08/an-epidemic-of-disbelief/592807/.

More on Carla Brooks can be found at Jennifer Weaver, "St. George Woman Forgives Rapist in Sentencing Hearing," 2KUTV, February 28, 2019, https:// kutv.com/news/local/st-george-elderly-woman-forgives-rapist-in-sentencing -hearing.

CHAPTER 25: THE NIETZSCHE DILEMMA

CeCe Moore's Mayflower Pilgrim post:

CeCe Moore, "Using Public Y-DNA Profiles to Track Down Criminals: Would You?," *Your Genetic Genealogist*, January 12, 2012, http://www.yourgenetic genealogist.com/2012/01/using-public-y-dna-profiles-to-track.html.

FamilyTreeDNA's Cooperation with Law Enforcement:

Salvador Hernandez, "One of the Biggest At-Home DNA Testing Companies Is Working with the FBI," *BuzzFeed News*, January 31, 2019, https://www .buzzfeednews.com/article/salvadorhernandez/family-tree-dna-fbi-investi gative-genealogy-privacy.

Matthew Haag, "FamilyTreeDNA Admits to Sharing Genetic Data with F.B.I.," *New York Times*, February 4, 2019, https://www.nytimes.com/2019 /02/04/business/family-tree-dna-fbi.html.

The *LA Times* on the Golden State Killer case:

Paige St. John, "The Untold Story of How the Golden State Killer Was Found: A Covert Operation and Private DNA," *Los Angeles Times*, December 8, 2020, https://www.latimes.com/california/story/2020-12-08/man-in-the-window.

CHAPTER 27: IS THAT IT?

N. F. Sugar, D. N. Fine, and L. O. Eckert, "Physical Injury after Sexual Assault: Findings of a Large Case Series," *American Journal of Obstetrics and Gynecology* 190, no. 1 (January 1, 2004): 71–76, https://doi.org/10.1016/s0002-9378(03) 00912-8.

Roy J. Levin and Willy van Berlo, "Sexual Arousal and Orgasm in Subjects Who Experience Forced or Non-Consensual Sexual Stimulation—A Review," *Journal of Clinical Forensic Medicine* 11, no. 2 (April 2004): 82–88, https://doi .org/10.1016/j.jcfm.2003.10.008.

INDEX

Note: Italicized page numbers indicate material in photographs or illustrations.

ABOUT THE AUTHOR

Edward Humes is a Pulitzer Prize–winning journalist and author whose fifteen previous books include *Burned*, *Mississippi Mud*, and the PEN award–winning *No Matter How Loud I Shout*. He became interested in the story of Tanya Van Cuylenborg and Jay Cook while living in Seattle, but now resides in Southern California with his human and greyhound family.